LAB WORKBOOK
Modern Masonry

Brick, Block, Stone

NINTH EDITION

Clois E. Kicklighter

Timothy L. Andera
Professor
South Dakota State University

Publisher
The Goodheart-Willcox Company, Inc.
Tinley Park, IL
www.g-w.com

Copyright © 2022
by
The Goodheart-Willcox Company, Inc.

All rights reserved. No part of this work may be reproduced, stored, or transmitted in any form or by any electronic or mechanical means, including information storage and retrieval systems, without the prior written permission of
The Goodheart-Willcox Company, Inc.

Manufactured in the United States of America.

ISBN 978-1-64564-666-2

3 4 5 6 7 8 9 – 22 – 25 24 23 22

The Goodheart-Willcox Company, Inc. Brand Disclaimer: Brand names, company names, and illustrations for products and services included in this text are provided for educational purposes only and do not represent or imply endorsement or recommendation by the author or the publisher.

The Goodheart-Willcox Company, Inc. Safety Notice: The reader is expressly advised to carefully read, understand, and apply all safety precautions and warnings described in this book or that might also be indicated in undertaking the activities and exercises described herein to minimize risk of personal injury or injury to others. Common sense and good judgment should also be exercised and applied to help avoid all potential hazards. The reader should always refer to the appropriate manufacturer's technical information, directions, and recommendations; then proceed with care to follow specific equipment operating instructions. The reader should understand these notices and cautions are not exhaustive.

The publisher makes no warranty or representation whatsoever, either expressed or implied, including but not limited to equipment, procedures, and applications described or referred to herein, their quality, performance, merchantability, or fitness for a particular purpose. The publisher assumes no responsibility for any changes, errors, or omissions in this book. The publisher specifically disclaims any liability whatsoever, including any direct, indirect, incidental, consequential, special, or exemplary damages resulting, in whole or in part, from the reader's use or reliance upon the information, instructions, procedures, warnings, cautions, applications, or other matter contained in this book. The publisher assumes no responsibility for the activities of the reader.

The Goodheart-Willcox Company, Inc. Internet Disclaimer: The Internet resources and listings in this Goodheart-Willcox Publisher product are provided solely as a convenience to you. These resources and listings were reviewed at the time of publication to provide you with accurate, safe, and appropriate information. Goodheart-Willcox Publisher has no control over the referenced websites and, due to the dynamic nature of the Internet, is not responsible or liable for the content, products, or performance of links to other websites or resources. Goodheart-Willcox Publisher makes no representation, either expressed or implied, regarding the content of these websites, and such references do not constitute an endorsement or recommendation of the information or content presented. It is your responsibility to take all protective measures to guard against inappropriate content, viruses, or other destructive elements.

Image Credit: Front cover: pics721/Shutterstock.com

Introduction

This lab workbook is intended to be used with the *Modern Masonry: Brick, Block, Stone* textbook. The lab workbook contains chapter review questions designed to reinforce the information presented in each chapter of the text, as well as jobs in which you apply your knowledge in hands-on activities.

Chapter Reviews

The lab workbook chapter reviews correspond to the textbook chapters and should be completed after reading the appropriate chapter in the text. The lab workbook chapter reviews are designed to enhance your understanding of textbook content. The various types of questions include multiple choice, true or false, matching, identification, and fill-in-the-blank. Answering the questions for each chapter will help you master the technical knowledge presented in the text.

Jobs

The lab workbook also contains jobs that cover areas specified in apprenticeship training programs. The jobs are action-oriented laboratory and field experiences that are designed to build skill and knowledge of accepted practices in the trade. The jobs are related to textbook content, and most of jobs are supported by information in the text in addition to the step-by-step procedures.

When performing the jobs, focus on your own safety and that of your coworkers. Follow all safety procedures as explained by your instructor. Ask questions if there is anything you do not understand.

Each job is a stand-alone activity that can be undertaken in any preferred sequence, but the jobs become more complex as the novice mason progresses through them in the recommended sequence. Also, more detailed information is given in the earlier jobs than in later ones. Each job adds to experience and lays the foundation for a more advanced experience in a related area.

As with any profession, there is more than one way to perform most tasks. Each job presents one method of performing the task, but other procedures may also be effective. Your instructor may modify a procedure.

Each job identifies the title or description of the job, the objective, the tools and equipment needed, and the recommended procedure for the job. Review questions at the end of each job are to be answered following the successful completion of the job.

Although not a specific requirement, apprentice masons should be encouraged to purchase their own equipment when possible. Using one's own equipment leads to better care and maintenance of it, which is also a necessary part of learning the trade. A list of the typical equipment for each trade area, as well as equipment that should be provided by the school, is included in the following section.

Mason's Tools

The following list of tools and equipment are presented as an aid to the apprentice mason. The basic tools listed in each trade area are representative of the tools most professionals use. The equipment is generally provided by the school for use in class activities.

Brick and Block Mason's Tools

Blocking chisel
Brick set chisel
Buttering trowel
Chalk box
Convex jointer (long)
Convex jointer (short)
Cross joint trowel
Duckbill trowel
Flexible rule
Folding rule
Framing square
Gauging trowel
Line blocks
Line trigs (twigs)
Margin trowel
Mason's brush
Mason's hammer
Mason's line
Mason's tool bag
Mason's trowel
Modular spacing rule
Personal safety equipment
Plain joint raker
Plugging or joint chisel
Plumb bob
Plumb rule
Pointing trowel
Skate wheel joint raker
Slicker
V-jointer (long)
V-jointer (short)

Stonemason's Tools

Buttering trowel
Chalk box
Convex jointer (long)
Convex jointer (short)
Cross joint trowel
Duckbill trowel
Flexible rule
Folding rule
Framing square
Gauging trowel
Line blocks
Margin trowel
Mason's brush
Mason's hammer
Mason's line
Mason's tool bag
Mason's trowel
Metal trigs (twigs)
Personal safety equipment
Plain joint raker
Plumb bob
Plumb rule
Pointing trowel
Pry bar
Skate wheel joint raker
Sledgehammer
Slicker
Sponge
Steel wedges
Stone mason's chisels
Stone mason's hammer
V-jointer (long)
V-jointer (short)

Cement Finisher's Tools

Bull float
Carpenter's hammer
Cement mason's hand trowel
Chalk box
Curb and gutter tools
Darby
Edgers (several sizes)
Flexible rule
Folding rule
Framing square
Hand float
Hand tamper
Jointers/groovers
Knee pads
Mason's hammer
Mason's line
Mason's trowel
Personal safety equipment
Plumb bob
Plumb rule
Screed
Step and corner tools
Tool bag
Wire cutters

Shop Tools and Equipment

Air compressor
Angle sander/grinder
Brick tongs
Buckets
Cement mixer
Concrete placer/spreader
Concrete rake
Concrete vibrator
Contractor's level and rod
Contractor's wheelbarrow
Corner poles
Dirt tamper
Forklift tractor
Grout bags
Hydraulic masonry splitter
Imprint roller
Jig saw
Ladders
Laser level
Masonry saw
Mixing box
Modular concrete forms
Mortar box
Mortar hoe
Mortar mixer
Mortar pans and stands
Portable circular saw
Portable concrete saw
Power activated nailer
Power compactor
Power elevator
Power float/troweler
Power screed
Rebar bender/cutter
Reciprocating saw
Rope and pulley
Rotary hammer
Round point shovel
Scaffolding
Square point shovel
Tuckpointer's grinder
Water hose
Wood mortar boards
Wooden body wheelbarrow

Contents

CHAPTER REVIEW

SECTION 1
Succeeding on the Job

Chapter 1 Careers in the Masonry Industry 1

Chapter 2 Job Performance, Leadership, Ethics, and Entrepreneurship 5

SECTION 2
Introduction to Masonry

Chapter 3 Tools and Equipment 7

Chapter 4 Safety 11

Chapter 5 Print Reading 15

Chapter 6 Math for Masonry Careers 19

SECTION 3
Materials

Chapter 7 Clay Masonry Materials 25

Chapter 8 Concrete Masonry Units 33

Chapter 9 Stone 39

Chapter 10 Mortar and Grout 43

Chapter 11 Anchors, Ties, and Reinforcement . . 49

SECTION 4
Techniques

Chapter 12 Laying Brick 53

Chapter 13 Laying Block 59

Chapter 14 Stonemasonry 63

SECTION 5
Construction Details

Chapter 15 Foundation Systems 69

Chapter 16 Wall Systems 73

Chapter 17 Paving and Masonry Construction Details 83

SECTION 6
Concrete

Chapter 18 Concrete Materials and Applications 87

Chapter 19 Form Construction 93

Chapter 20 Concrete Flatwork and Formed Shapes 99

JOBS

- **Job 1** Identifying Common Masonry Materials ... 107
- **Job 2** Identifying Types of Clay Brick ... 111
- **Job 3** Arranging Brick in the Five Basic Structural Bonds ... 115
- **Job 4** Identifying Common Concrete Masonry Units ... 119
- **Job 5** Identifying Common Concrete Block Shapes by Name ... 123
- **Job 6** Identifying Common Building Stone Samples ... 127
- **Job 7** Measuring Mortar Materials and Mixing Mortar ... 131
- **Job 8** Identifying Anchors, Ties, and Joint Reinforcement ... 133
- **Job 9** Loading the Trowel and Spreading a Mortar Bed for Brick and Block ... 135
- **Job 10** Forming a Head Joint on Brick and Block ... 139
- **Job 11** Cutting Brick with a Brick Hammer ... 143
- **Job 12** Cutting Brick with a Brick Set Chisel ... 145
- **Job 13** Cutting Brick with a Mason's Trowel ... 147
- **Job 14** Cutting Brick with a Masonry Saw ... 149
- **Job 15** Cutting Concrete Block with a Brick Hammer and Blocking Chisel ... 151
- **Job 16** Using a Mason's Line ... 153
- **Job 17** Erecting Batter Boards ... 155
- **Job 18** Laying Six Bricks on a Board ... 157
- **Job 19** Laying a Four Course, Single Wythe, Running Bond Lead ... 161
- **Job 20** Laying a 4" Running Bond Wall with Leads ... 165
- **Job 21** Laying an 8" Common Bond, Double Wythe Brick Wall with Leads ... 169
- **Job 22** Laying an 8" Running Bond, Two-Wythe Intersecting Brick Wall ... 173
- **Job 23** Building a Corner in Flemish Bond with Quarter Closures ... 177
- **Job 24** Building a Corner in Flemish Bond with Three-Quarter Closures ... 179
- **Job 25** Building a Corner in English Bond with Quarter Closures ... 181
- **Job 26** Building a Corner in English Bond with Three-Quarter Closures ... 183
- **Job 27** Constructing a 10" Brick Masonry Cavity Wall with Metal Ties and Weep Holes ... 185
- **Job 28** Constructing a Reinforced Single Wythe Brick Bearing Wall ... 189
- **Job 29** Corbeling a 12" Brick Wall ... 191
- **Job 30** Building a Hollow Brick Masonry Pier ... 193
- **Job 31** Cleaning New, Dark Colored Brick Masonry ... 197
- **Job 32** Acid Cleaning of Light Colored Brick ... 199
- **Job 33** Handling Concrete Blocks ... 201
- **Job 34** Laying an 8" Running Bond Concrete Block Wall ... 205
- **Job 35** Laying a 10" Concrete Block Cavity Wall ... 209
- **Job 36** Laying an 8" Composite Wall with Concrete Block Backup ... 213
- **Job 37** Cleaning Concrete Block Masonry ... 215
- **Job 38** Handling Stone ... 217
- **Job 39** Forming Mortar Joints in Stone Masonry ... 221
- **Job 40** Splitting, Shaping, and Cutting Stone ... 223
- **Job 41** Setting a Random Rubble Stone Veneer Wall ... 225

Job 42 Building a Solid 12″ Thick Ashlar Stone Wall 229	**Job 61** Building a Hollow Brick Masonry Bonded Wall 283
Job 43 Setting a Limestone Panel 231	**Job 62** Building an Anchored Veneered Wall and Installing Flashing 285
Job 44 Pointing Cut Stone after Setting 233	**Job 63** Building a 12″ Composite Brick and Block Wall 289
Job 45 Cleaning New Stone Masonry 235	**Job 64** Building Reinforced Masonry Walls . . . 291
Job 46 Applying Manufactured Stone to a Backup 237	**Job 65** Installing Steel and Concrete Reinforced Lintels 295
Job 47 Measuring Concrete Materials 241	
Job 48 Mixing Concrete with a Power Mixer . . . 243	**Job 66** Building Reinforced Concrete Block and Brick Lintels 297
Job 49 Performing a Slump Test on Plastic Concrete. 245	**Job 67** Building Masonry Sills and Installing Stone Sills 301
Job 50 Placing Concrete in a Slab Form 247	**Job 68** Building a Brick Masonry Arch 305
Job 51 Finishing Concrete Slabs 249	**Job 69** Forming Movement Joints in Concrete and Masonry 309
Job 52 Building Footing Forms for Concrete. 253	**Job 70** Installing Masonry Pavers on a Rigid Base. 313
Job 53 Building Wall Forms for Concrete 257	**Job 71** Building Concrete and Masonry Steps. . 315
Job 54 Building and Installing a Buck 261	**Job 72** Building a Masonry Fireplace and Chimney 317
Job 55 Installing Round Column Forms 263	
Job 56 Building Centering for a Masonry Arch. 265	**Job 73** Building a Masonry Garden Wall with Coping. 321
Job 57 Dampproofing Concrete Block Basement Walls 269	**Job 74** Corbeling and Racking a Masonry Wall. 323
Job 58 Building Columns, Piers, and Pilasters 273	**Job 75** Building a Mortarless Retaining Wall . . 325
Job 59 Building Solid Masonry Walls 277	**Job 76** Using a Corner Pole 327
Job 60 Building a 4″ RBM Curtain or Panel Wall 281	

Name _____ Date _____ Class _____

CHAPTER 1
Careers in the Masonry Industry

Carefully read Chapter 1 of the text and answer the following questions.

1. _____ Good health and _____ are important qualities for masons to have.
    A. strength
    B. eyesight
    C. manual dexterity
    D. All of the above.

2. _____ Masons must be able to estimate _____.
    A. volumes
    B. weights
    C. quantities
    D. All of the above.

3. _____ The term _____ has traditionally referred to the craft of building with bricks.
    A. construction
    B. masonry
    C. carpentry
    D. blocking

4. _____ Cement masons work with _____.
    A. brick
    B. concrete
    C. stone
    D. tile

5. _____ Classes in _____ and blueprint reading help to develop knowledge of layout and design.
    A. drafting
    B. industrial technology
    C. woodworking
    D. carpentry

6. _____ _____ programs allow students to continue their learning experiences at an employer's worksite during a designated portion of the school day.
    A. SkillsUSA
    B. Internship
    C. Dual-learning
    D. Job Corps

7. _____ _____ promotes leadership and personal growth and conducts career competitions.
    A. Associated General Contractors of America
    B. National Association of Home Builders
    C. SkillsUSA
    D. Job Corps

8. _____ *True or False?* An individual may become a mason tender as part of an apprenticeship program, but this cannot lead into a masonry apprenticeship.

9. _____ The minimum age for an apprentice is _____ years old.
    A. 21
    B. 18
    C. 16
    D. There is no minimum age limit.

Copyright Goodheart-Willcox Co., Inc.
May not be reproduced or posted to a publicly accessible website.

10. _____ A person who signs an agreement to work and learn a trade is called a(n) _____.
    A. apprentice
    B. foreman
    C. journeyman
    D. masonry student

11. _____ The normal term of apprenticeship varies, but is usually about _____ years.
    A. two
    B. three
    C. four
    D. five

12. _____ Apprenticeship is divided into six periods of advancement at _____ months each.
    A. three
    B. six
    C. nine
    D. twelve

13. _____ Upon successful completion of the apprenticeship program, the apprentice becomes a _____.
    A. contractor
    B. foreman
    C. journeyman
    D. project manager

14. _____ A journeyman who has the responsibility of supervising a group of workers is called a _____.
    A. project manager
    B. supervisor
    C. contractor
    D. foreman

15. _____ The person who works under the direction of a superintendent to keep a project on time and on budget is called a _____.
    A. contractor
    B. foreman
    C. journeyman
    D. project manager

16. _____ The _____ supervises the work of the subcontractors, project manager, and foremen.
    A. supervisor
    B. contractor
    C. superintendent
    D. foreman

17. _____ The goal of networking is to _____.
    A. gain an apprenticeship
    B. improve masonry skills
    C. learn about possible job leads
    D. become a superintendent

18. _____ A résumé should contain all of the following, *except* _____.
    A. present job position
    B. educational background
    C. professional accomplishments
    D. a list of references

19. _____ *True or False?* Along with a résumé, a job applicant should develop a separate list of at least five references.

Name _____

20. _____ Which of the following should *not* be used as a reference?
    A. School official
    B. Relative
    C. Previous employer
    D. Teacher

21. _____ *True or False?* When you send a résumé to a potential employer, a letter of application should be sent with it.

22. _____ Which of the following should *not* be included in the letter of application?
    A. Title of the job you seek
    B. Where you heard about the job
    C. List of previous jobs held
    D. Request for an interview

23. _____ *True or False?* When filling out a job application, you should leave a section blank if it asks for information that does not apply to you.

24. _____ *True or False?* When filling out an online application, include keywords for which the employer may search.

25. _____ When preparing for a job interview, you should _____.
    A. ask about salary compensation
    B. learn about the company
    C. list the vacation time you will need
    D. research the educational level of the interviewer

26. _____ *True or False?* During a job interview, it is important to answer all questions as carefully and completely as you can.

27. _____ When considering a job offer, which of the following factors should be considered?
    A. Work schedule
    B. Job obligations
    C. Income benefits
    D. All of the above.

Notes

Name _____ Date _____ Class _____

CHAPTER 2: Job Performance, Leadership, Ethics, and Entrepreneurship

Carefully read Chapter 2 of the text and answer the following questions.

1. _____ Three primary areas that are considered when evaluating job performance are general work habits, _____ record on the job, and the ability to stay current in the field.
   A. on-time
   B. safety
   C. personal
   D. None of the above.

2. _____ A cluttered workplace encourages _____.
   A. teamwork
   B. efficiency
   C. accidents
   D. hard work

3. _____ Which of the following is *not* a quality employers generally look for in employees?
   A. Punctuality
   B. Dependability
   C. Friendliness
   D. Responsibility

4. _____ Taking initiative means _____.
   A. always being on time
   B. observing company policies
   C. starting activities without being told
   D. being accurate and error-free

5. _____ *True or False?* Constant attention to safety guidelines is crucial to preventing injuries and safety violations.

6. _____ Workplace accidents occur because of _____.
   A. carelessness
   B. disobeying company rules
   C. poor attitudes
   D. All of the above.

7. _____ *True or False?* When a person moves beyond the level of apprenticeship, the educational process stops.

8. _____ Your _____ includes how you handle job responsibilities, including your enthusiasm for work.
   A. work experience
   B. work ethic
   C. teamwork ability
   D. skill

9. _____ *True or False?* Your work ethic includes how you relate to fellow workers and supervisors.

10. _____ *True or False?* On the job, each worker should always behave as if he or she were in charge.

11. Leadership is the ability to guide and _____ others to complete tasks or achieve goals.

12. _____ Good leaders _____.
    A. can get along with all types of people
    B. are likely to be promoted to higher levels
    C. meet problems head-on
    D. All of the above.

13. Good leaders lead by _____, meaning that they practice the types of behavior they expect from others.
    _____

14. _____ Having the ability to solve problems on the job shows an employer that you are able to handle more _____.
    A. pay
    B. responsibility
    C. important work
    D. All of the above.

15. _____ *True or False?* Solving problems as a group can strengthen a team.

16. Decision-making and problem-solving require _____ thinking skills.
    _____

17. _____ Someone who organizes, manages, and assumes the risks of a business is called a(n) _____.
    A. entrepreneur
    B. contractor
    C. project manager
    D. superintendent

18. _____ *True or False?* A good idea or quality product or service is enough to launch and sustain a successful business.

19. _____ To make a profit, an entrepreneur must know all aspects of _____.
    A. business
    B. the trade
    C. consumer habits
    D. the housing market

20. _____ Ethics can be defined as "the rules or standards governing the _____ of members of a profession."
    A. livelihood
    B. attitude
    C. conduct
    D. dependability

21. _____ *True or False?* Ethical practice is a concern of businesses that wish to be successful over the long term.

22. List six characteristics of a successful entrepreneur.
    _____
    _____
    _____

Name _____ Date _____ Class _____

CHAPTER 3
Tools and Equipment

Carefully read Chapter 3 of the text and answer the following questions.

1. _____ The tool most commonly used by a mason is the _____.
    A. brick hammer
    B. mason's trowel
    C. mortar hoe
    D. jointer

2. _____ Each of the following is a trowel style, *except* _____.
    A. London pattern
    B. Philadelphia pattern
    C. Wide London pattern
    D. Wide Philadelphia pattern

3. _____ The metal band found around the shank end of a wooden-handled trowel is called a _____.
    A. ferrule
    B. point
    C. heel
    D. blade

4. _____ A _____ trowel is used for mixing mortar or other masonry products.
    A. duck bill or coke
    B. gauging
    C. margin
    D. tuck pointer

5. _____ A _____ trowel is used for filling or shaping mortar between brick.
    A. gauging
    B. pointing
    C. margin
    D. tuck pointer

6. _____ The mason's tool used to finish the surface of mortar joints is called a _____.
    A. brick hammer
    B. jointer
    C. level
    D. trowel

7. _____ *True or False?* A chisel hammer is a hammer designed to drive nails and break or chip stone or masonry units.

8. _____ Mortar boxes are typically supplied by the _____.
    A. supplier
    B. contractor
    C. foreman
    D. superintendent

9. _____ A mortar _____ is used to mix mortar in a mortar box or mason's wheelbarrow.
    A. hoe
    B. bag
    C. stand
    D. board

10. _____ The purpose of a mortar _____ is to hold mortar during the time a mason is laying brick or concrete block.
    A. stand
    B. board
    C. box
    D. hoe

11. _____ The type of rule used depends on the specific project being built and the _____.
    A. quality of the rule
    B. skill of the worker using it
    C. layout requirements
    D. None of the above.

12. _____ A steel square has a tongue that is 1 1/2″ wide and 16″ long, with a body that is 2″ wide and _____ long.
    A. 20″
    B. 22″
    C. 24″
    D. 26″

13. _____ A level used by masons to check a wall to ensure that it is built vertical and level is called a mason's level or a _____ rule.
    A. mason's
    B. plumb
    C. level
    D. steel

14. _____ *True or False?* Electronic and rotary laser levels are both extremely accurate.

15. _____ *True or False?* When striking chisels, it is best to use a brick hammer.

16. _____ The strong nylon cord that is used to keep each course level and the wall "true" and "out-of-wind" is called a mason's _____.
    A. cable
    B. line
    C. level
    D. thread

17. _____ _____ are hooked onto each corner of a line to aid a mason in laying bricks or concrete blocks at the correct height.
    A. Line pins
    B. Twigs
    C. Brick tongs
    D. Line blocks

18. _____ *True or False?* Masonry brushes and sponges can be used for cleaning brick or placing a finish texture on a masonry surface.

19. _____ _____ are used to carry brick.
    A. Brick tongs
    B. Brackets
    C. Twigs
    D. Wheelbarrows

Name _____

20. Review the following image and identify the parts of the scaffolding.

   A. _____
   B. _____
   C. _____
   D. _____
   E. _____
   F. _____
   G. _____
   H. _____
   I. _____

   Goodheart-Willcox Publisher

21. _____ A typical mortar mixer mixes about _____ cu ft of mortar at a time.
   A. 2
   B. 4
   C. 6
   D. 8

22. _____ A _____ cuts masonry units faster and more accurately than a mason's chisel and hammer.
   A. mason's saw
   B. hand tamper
   C. tamping rammer compactor
   D. vibratory plate compactor

23. _____ The primary use of a _____ is to prep subbases for footings, driveways, sidewalks, and basement floors.
   A. hand tamper
   B. tamping rammer compactor
   C. vibratory plate connector
   D. track-driven power buggy

24. _____ A track-driven power buggy can carry _____ cu ft of material.
   A. 4
   B. 8
   C. 14
   D. 20

25. _____ The first finishing tool used by a cement mason after the concrete is placed is called a _____.
   A. bull float
   B. darby
   C. screed
   D. tamper

26. _____ A tool used by a cement mason to compact concrete into a dense mass is called a _____.
   A. hand tamper
   B. screed
   C. darby
   D. bull float

27. _____ *True or False?* A darby is used to remove high or low spots left by the screed.

28. A large, flat, rectangular piece of wood, aluminum, or magnesium that allows masons to float a large area is called a(n) _____ float.

_____

29. _____ Which of the following tools is used to produce a radius on the edge of a slab?
    A. Bull float
    B. Darby
    C. Edger
    D. Groover

30. _____ *True or False?* Concrete brooms are used to cut into the surface of freshly placed concrete.

31. A tool that is used to cut joints in concrete slabs is called a concrete _____ saw.

_____

32. _____ Hand _____ are used to prepare concrete surfaces for troweling.
    A. brooms
    B. edgers
    C. floats
    D. tampers

33. _____ A _____ moves or oscillates to remove air pockets or voids in concrete placed in walls, floors, footings, and other concrete forms.
    A. concrete vibrator
    B. power float
    C. power trowel
    D. tamping rammer compactor

34. _____ *True or False?* A concrete mason's hand finishing trowel is the last tool used in the finishing process of a concrete slab.

Name _____ Date _____ Class _____

CHAPTER 4: Safety

Carefully read Chapter 4 of the text and answer the following questions.

1. _____ *True or False?* Construction is cited by the Occupational Safety and Health Administration as one of the safest occupations in the United States.

2. _____ Which of the following is the abbreviation for the organization responsible for developing standards to ensure safe and healthy working conditions for construction workers?
   A. MSDS
   B. OSHA
   C. PPE
   D. US DOT

3. _____ What is the first step of the Job Hazard Analysis (JHA) process?
   A. Identifying ways to engineer hazards out a building.
   B. Foreman or supervisor does a walk-through and discusses hazards found.
   C. Safety training is conducted for new employees.
   D. The hazards are eliminated.

4. _____ As prescribed by OSHA, _____ safety training should be conducted.
   A. daily
   B. monthly
   C. weekly
   D. yearly

5. _____ To keep safety uppermost in the minds of all workers, short safety talks called _____ talks are given by a foreman or supervisor.
   A. safety
   B. toolbox
   C. walk-through
   D. hazard

6. _____ *True or False?* Safety meetings need to be held only at the start of a construction project.

7. _____ *True or False?* Learning safe work practices is as much a part of learning a trade as using the tools.

8. _____ The personal protective equipment that is worn by workers on building and highway construction sites so the wearer can be more readily seen is called a _____ vest.
   A. high-visibility
   B. construction
   C. building
   D. PPE

9. _____ Many construction industries in the masonry trade require everyone on a jobsite to wear _____ protection 100% of the time.
   A. fall
   B. head
   C. hand
   D. ear

10. _____ *True or False?* Keeping tools and equipment in good repair and using them properly is the best way to avoid accidents with them.

11. _____ In addition to eye and ear protection, a(n) _____ should be worn when using a power mixer to mix dry materials or concrete products.
    A. hard hat
    B. HPPE glove
    C. face shield
    D. respirator

12. _____ If an electrical accident occurs and someone may have a live current flowing through him or her, the first step to take, if possible, is to _____.
    A. break the contact with a dry piece of wood
    B. call for assistance
    C. pull the individual away from the source of electricity
    D. shut off the power

13. _____ The process of locking the machine or the electrical power out so the equipment will not operate is called _____.
    A. boltout
    B. lockout
    C. markout
    D. tagout

14. OSHA requires all temporary power on a construction site to incorporate a(n) _____ as a part of the electrical system.
    _____

15. _____ Which of the following is *not* one of the important rules to follow when handling materials?
    A. Do not exceed maximum stacking heights.
    B. Do not overload a wheelbarrow.
    C. Lift heavy loads using your back.
    D. Store materials on a paved surface.

16. _____ OSHA's Construction Scaffold Standard specifies that the erection of a scaffold must be directed by a _____ person to ensure that it is erected according to industry practices.
    A. competent
    B. qualified
    C. trained
    D. All of the above.

17. _____ OSHA requires that construction workers be protected when working at levels above _____ in most types of construction activities.
    A. 5′
    B. 6′
    C. 7′
    D. 8′

18. _____ *True or False?* Safety nets are placed in elevator shafts, between floors, or under rafters to catch a falling worker.

19. _____ What fuel is used to heat most temporary heaters for enclosures?
    A. propane
    B. natural gas
    C. kerosene
    D. All of the above.

20. _____ Which of the following is *not* a correct ladder safety rule?
    A. The area around the bottom of a ladder should be clear of clutter.
    B. Be sure the ladder extends at least 1′ above the point where you plan to step off it.
    C. Keep ladders out of walkways and traffic lanes.
    D. Never stand on the highest rung of a ladder.

21. _____ *True or False?* Increased exposure to silica dust may be fatal.

22. A(n) _____ vacuum is required to vacuum in areas that may contain silica.
    _____

Name _____

23. _____ The written documents that outline product information and procedures to be used when working with or handling a product or chemical are called _____.
    A. material data sheets
    B. safety data sheets
    C. hazard statements
    D. OSHA data sheets

24. _____ Which of the following signal words would be located under a pictogram frame to indicate severe hazards?
    A. Caution
    B. Danger
    C. Hazard
    D. Warning

25. _____ When using chemicals, the two standard PPE items required are rubber gloves and _____.
    A. face shields
    B. respirators
    C. hard hats
    D. safety glasses

26. _____ is an emergency technique used on individuals when the heart has stopped beating.
    _____

27. List three basic first aid rules.
    _____
    _____
    _____

Notes

Name _____ Date _____ Class _____

CHAPTER 5
Print Reading

Carefully read Chapter 5 of the text and answer the following questions.

1. _____ *True or False?* Construction drawings describe the size, shape, location, and specification of the elements of a structure.

2. _____ Construction drawings may be called all of the following, *except* _____.
   A. blueprints
   B. plans
   C. sketches
   D. working drawings

3. _____ The different types of lines used on construction drawings are called the _____ of Lines.
   A. Symbol
   B. Alphabet
   C. Blueprint
   D. None of the above.

4. _____ The very heavy lines that are drawn to form a boundary for a construction drawing are called _____ lines.
   A. border
   B. object
   C. center
   D. extension

5. _____ *True or False?* Border lines are heavy lines that show the outline of the visible features of an object.

6. _____ Which of the following types of lines are medium-weight lines that represent an edge or intersection of two surfaces that are *not* visible in a given view?
   A. Border lines
   B. Center lines
   C. Extension lines
   D. Hidden lines

Match each construction drawing line type with its definition.

7. _____ Heavy lines used to indicate where the object has been sectioned to show internal features.

8. _____ Heavy lines used when part of the object is shown broken away to reveal a hidden feature.

9. _____ Thin lines used to show size or location of a feature of the structure.

10. _____ Thin lines used to show that all of the part is not shown.

11. _____ Thin lines used to show where a dimension line ends.

A. Dimension line
B. Cutting plane line
C. Extension line
D. Long break line
E. Short break line

12. _____ Section lines are very thin lines used to show that the feature has been sectioned; these are also called _____ lines.
    A. crosshatch
    B. portion
    C. segment
    D. traverse

13. _____ The purpose of symbols and abbreviations is to communicate efficiently and conserve _____.
    A. time
    B. space
    C. materials
    D. effort

14. _____ Residential floor plans, elevations, and foundation plans are generally drawn at a _____ scale.
    A. 1/8″ = 1′-0″
    B. 1/6″ = 1′-0″
    C. 1/4″ = 1′-0″
    D. 1/2″ = 1′-0″

15. _____ *True or False?* On a residential construction drawing, details are usually drawn at a smaller scale than the rest of the drawing.

16. _____ *True or False?* Dimensions indicate the size and location of building elements and must be followed accurately.

17. _____ Interior wood frame walls may be dimensioned in any of the following ways, *except* _____.
    A. to the center of the wall
    B. to the inside of the studs
    C. to the outside of the finished wall
    D. to the outside of the studs

18. _____ Interior and exterior masonry or concrete walls are dimensioned to the _____ of the walls.
    A. outside
    B. inside
    C. center
    D. All of the above.

19. _____ Metric units of measure in the construction industry should be restricted to the _____ and the _____.
    A. inch; foot
    B. meter; millimeter
    C. yard; mile
    D. millimeter; kilometer

20. _____ Which of the following is the correct term for the drawings used to construct a building?
    A. Architectural renderings
    B. Presentation drawings
    C. Preliminary drawings
    D. Working drawings

21. _____ Corrections and revisions are recorded on the _____ drawings and form the basis for the working drawings.
    A. presentation
    B. preliminary
    C. perspective
    D. architectural

22. _____ *True or False?* Presentation drawings are generally picture-like drawings that show how the finished structure will appear.

23. _____ The type of pictorial drawing most often used by architects is the _____ drawing.
    A. graphic
    B. perspective
    C. picturesque
    D. prospect

24. _____ The working drawing that shows the location of the building and may also show utilities, topographical features, streets, walks, and other items is called the _____ plan.
    A. site
    B. foundation
    C. floor
    D. mechanical

Name _____

25. _____ A(n) _____ plan shows all exterior and interior walls, doors, windows, patios, walks, decks, fireplaces, built-in cabinets, and appliances.
    A. construction
    B. electrical
    C. floor
    D. foundation

26. _____ Which of the following drawings shows outside features, such as placement and height of windows, doors, chimney, and roof lines?
    A. Foundation
    B. Elevation
    C. Floor
    D. All of the above.

27. _____ *True or False?* The electrical plan locates switches, convenience outlets, ceiling outlet fixtures, TV jacks, service entrance location, and panel box.

28. _____ A mechanical plan shows the plumbing, heating, and cooling systems. What is another name for this type of plan?
    A. Shop drawing
    B. Climate control plan
    C. Framing plan
    D. Environment plan

29. _____ Plans that provide construction details of specific areas of a building are called _____ drawings.
    A. presentation
    B. shop
    C. elevation
    D. detail

30. _____ *True or False?* The foundation plan is designed to show in detail the framing requirements for a roof, floor, or other framed area.

31. _____ A plan that locates and identifies trees, plants, and hardscape is called the _____ plan.
    A. roof
    B. framing
    C. landscaping
    D. floor

32. _____ Drawings that provide additional information about reinforcing steel, complex cabinetwork, and electronic systems are called _____ drawings.
    A. data
    B. fact
    C. shop
    D. updated

33. _____ *True or False?* Shop notes are included in a set of construction drawings to clarify details that might otherwise cause confusion.

34. _____ An agreement is made between all parties involved in a project based on construction documents that include working drawings and _____.
    A. architectural plans
    B. specifications
    C. work schedule
    D. None of the above.

35. _____ What is the section of specifications that deals with items such as insurance and responsibility for permits?
    A. Common conditions
    B. General conditions
    C. Technical sections
    D. None of the above.

36. _____ *True or False?* Technical sections are specifications that deal with the actual building of the structure and the tradework involved in the building.

37. Review the following images and identify the different types of lines used on construction drawings.

A. ______________________________

B. ______________________________

C. ______________________________

D. ______________________________

E. ______________________________

F. ______________________________

G. ______________________________

H. ______________________________

I. ______________________________

Goodheart-Willcox Publisher

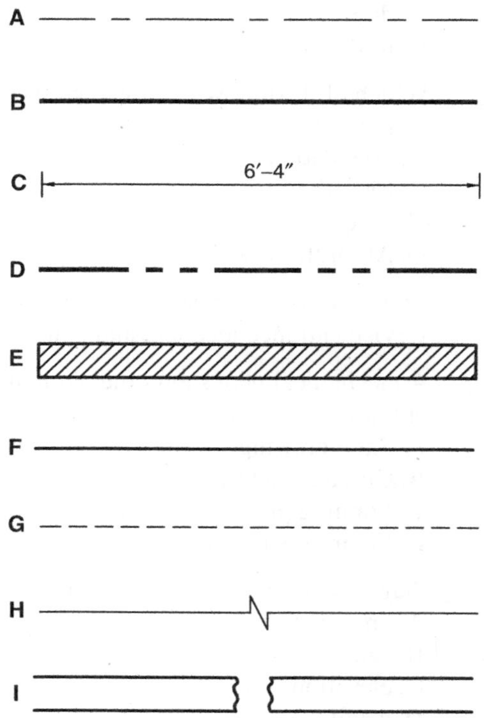

Name _____ Date _____ Class _____

CHAPTER 6
Math for Masonry Careers

Carefully read Chapter 6 of the text and answer the following questions.

1. _____ *True or False?* A whole number is any one of the natural numbers, such as 1, 2, or 5.

2. _____ The process of combining two or more numbers to find a total is called _____.
   A. addition
   B. division
   C. multiplication
   D. subtraction

3. The total in addition is called the _____.

_____

4. _____ Weight slips on four loads of gravel delivered to one job were 6272 lb, 6098 lb, 6138 lb, and 6193 lb. The total weight of the gravel delivered to the job was _____.
   A. 24,602 lb
   B. 24,656 lb
   C. 24,701 lb
   D. 24,746 lb

5. _____ A mason contracts a job for $700. Labor cost is $316 and material cost is $203. Other expenses total $78. How much profit is made on the job?
   A. $103
   B. $181
   C. $597
   D. $622

6. If 6 + 6 + 6 + 6 + 6 need to be added, the shortest method would be to multiply _____ times 6 to get a product of 30.

_____

7. _____ A bricklayer lays 160 bricks per hour. At this rate, how many bricks will he lay in seven hours?
   A. 1160
   B. 1120
   C. 1000
   D. 1610

8. _____ The answer to a division problem is the _____.
   A. product
   B. divisor
   C. quotient
   D. dividend

9. _____ Three bricklayers laid 3150 bricks on a job in one day. What is the average number of bricks laid by each bricklayer that day?
   A. 50
   B. 525
   C. 1050
   D. None of the above.

10. _____ The top number of a fraction is called the _____.
    A. denominator
    B. dividend
    C. divisor
    D. numerator

11. _____ A whole number followed by a fraction is called a(n) _____ number.
    A. improper
    B. mixed
    C. decimal
    D. split

12. _____ The easiest fractions to add are those that have the same _____, such as 1/8 and 3/8.
    A. numerators
    B. denominators
    C. divisors
    D. quotients

13. _____ The procedure for multiplying fractions is to multiply the denominators together and then multiply the _____ together to find the answer.
    A. mixed numbers
    B. numerators
    C. product of the denominators
    D. difference

14. _____ If 2 1/4" bricks are used to lay a wall with 3/8" mortar joints, what is the height of the wall after seven courses?
    A. 18 3/8"
    B. 18 1/8"
    C. 14 3/8"
    D. 17 5/8"

15. _____ *True or False?* Fractions are divided by inverting the divisor and then multiplying.

16. _____ *True or False?* Decimals are types of fractions that have only the denominator.

17. _____ The decimal 0.025 is equal to _____.
    A. 25/100
    B. 25/1000
    C. 25/10,000
    D. 25/100,000

18. _____ A decimal _____ number has numbers on both sides of the decimal.
    A. mixed
    B. improper
    C. whole
    D. fraction

19. _____ If the cost of materials for a masonry job is $35.50 for sand, $162.38 for cement, $9.63 for one steel lintel, and $338.19 for brick, what is the total cost of materials for the job?
    A. $545.70
    B. $559.70
    C. $590.70
    D. $595.70

20. _____ What is the cost of 22 bags of cement if each bag costs $3.50?
    A. $6.29
    B. $77.00
    C. $138.29
    D. $144.00

21. _____ *True or False?* Common fractions can be changed to decimals by dividing the numerator by the denominator.

Name _____

22. _____ Reinforcing steel required for a certain situation is .375″ in diameter. What is the fractional size of this steel?
    A. 1/4″
    B. 3/8″
    C. 1/2″
    D. 5/8″

23. _____ Percentage is a number used to denote parts of _____.
    A. 10
    B. 50
    C. 100
    D. 1000

24. _____ The cost of borrowing money is called _____.
    A. principal
    B. loan
    C. interest
    D. rate of interest

25. _____ *True or False?* The amount of money borrowed in a loan is the rate of interest.

26. _____ If the annual interest on a loan is 6%, the amount of interest owed at the end of a year on a $7000 loan would be _____.
    A. $116.67
    B. $420
    C. $1166.67
    D. $7420

27. _____ Basic elements of lines that have no width, length, or height, and merely indicate a location are called _____.
    A. points
    B. circles
    C. geometric shapes
    D. angles

28. _____ A(n) _____ has only length and is the distance between two or more points.
    A. point
    B. line
    C. angle
    D. circle

29. _____ Perpendicular means that the line forms a _____ angle with respect to another line.
    A. 45°
    B. 60°
    C. 90°
    D. 180°

30. _____ Lines that remain a constant distance apart and never cross are called _____ lines.
    A. perpendicular
    B. parallel
    C. diagonal
    D. None of the above.

Match the name of the angle with the correct degree or definition.

31. _____ 90°

32. _____ 180°

33. _____ Less than 90°

34. _____ Greater than 90°

A. Acute angle
B. Obtuse angle
C. Right angle
D. Straight angle

35. Review the following image and identify the parts of the circle.

   A. _____

   B. _____

   C. _____

   D. _____

   E. _____

   F. _____

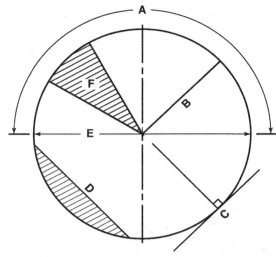
Goodheart-Willcox Publisher

36. _____ The distance around a circle is referred to as its _____.
   A. diameter
   B. radius
   C. circumference
   D. arc

37. _____ What is the formula for finding the circumference of a circle?
   A. $A^2 + B^2 = C^2$
   B. C = 3.1416 × diameter
   C. C = πD
   D. Both B and C.

38. _____ A triangle is a geometric shape with three angles that total _____.
   A. 45°
   B. 90°
   C. 120°
   D. 180°

39. _____ *True or False?* The hypotenuse is on the same side as the 90° angle of a right triangle.

40. _____ If the length of the sides of a right triangle have a 3:4:5 proportion, the triangle is a(n) _____ triangle.
   A. acute
   B. obtuse
   C. right
   D. true

41. _____ *True or False?* Areas are two-dimensional quantities (length × width) and are shown in square measure.

42. _____ Volumes are three-dimensional quantities (length × width × height) that are shown in _____ measure.
   A. square
   B. cubic
   C. 3-D
   D. metric

43. Relabeling from the US customary to the metric system without physically changing the product is called _____ metric conversion.

_____

Name _____

44. Review the following chart of base units and fill in the correct SI unit for each quantity.

   A. _____
   B. _____
   C. _____
   D. _____
   E. _____
   F. _____
   G. _____

Quantity	SI Unit	SI Symbol
Length	A	m
Mass (weight)	B	kg
Time	C	s
Temperature	D	K
Electric current	E	A
Luminous intensity	F	cd
Amount of substance	G	mol

45. _____ The units that are formed by multiplication or division of two or more SI base units are called _____ units.
   A. supplemental
   B. non-SI
   C. derived
   D. All of the above.

46. _____ SI prefixes are formed by multiplying or dividing base units by powers of _____.
   A. 10
   B. 20
   C. 100
   D. None of the above.

47. _____ Ten inches are equal to _____.
   A. 2.54 centimeters
   B. 25.4 centimeters
   C. 30.48 centimeters
   D. 2.54 meters

48. _____ The US customary to metric conversion for lengths is 1 ft = 30.48 cm. What are the dimensions in meters of a residential structure that is 40′ × 60′?
   A. 1.312 m × 1.969 m
   B. 12.191 m × 18.288 m
   C. 76.2 m × 50.8 m
   D. 131.24 m × 229.26 m

49. _____ Which of the following is the most widely used procedure for estimating brick masonry?
   A. Wall-area method
   B. Assessment method
   C. Section method
   D. 3-4-5 method

50. The gross area less (minus) all wall openings is referred to as the _____ wall area.

_____

51. _____ As a general rule of thumb, the mortar yield is _____ cu ft for each 1 cu ft of damp loose sand.
   A. 1/2
   B. 1
   C. 2
   D. 3

Notes

Name _____ Date _____ Class _____

CHAPTER 7
Clay Masonry Materials

Carefully read Chapter 7 of the text and answer the following questions.

1. The _____ and the ASCE have adopted a consensus standard for the design of masonry structures and standard construction specifications.

   _____

2. _____ What year was the International Code Council (ICC) formed?
   A. 1988
   B. 1989
   C. 1994
   D. 1995

3. _____ *True or False?* The International Residential Code is an I-Code.

4. _____ What is the name for the joint effort between different industry organizations to create codes that cover the design and construction of masonry structures?
   A. American Concrete Institute (ACI)
   B. American Society of Civil Engineers (ASCE)
   C. Masonry Standards Joint Committee (MSJC)
   D. The Masonry Society (TMS)

5. _____ Structural facing tile is classified as _____.
   A. a hollow masonry unit
   B. a solid masonry unit
   C. architectural terra cotta
   D. brick

6. _____ A crystalline mineral made up of aluminum silicates with either potassium, sodium, calcium, or barium is called _____.
   A. shale
   B. feldspar
   C. flint
   D. silica

7. _____ *True or False?* In the water-struck method of the soft mud process, the brick usually has a sandpaper-like surface.

8. _____ Which of the following methods of manufacturing bricks involves the clay being forced through a die and extruded in the form of a ribbon or column?
   A. Dry-press process
   B. Hard mud process
   C. Soft mud process
   D. Stiff mud process

9. _____ Bricks that have not been fired are called _____ bricks.
   A. green
   B. dried
   C. loose
   D. wet

10. Two types of kilns used for drying bricks are periodic kilns and _____ kilns.

    _____

11. _____ Building brick is often referred to as _____ brick.
    A. facing
    B. standard
    C. common
    D. hollow

12. _____ Which of the following building brick types is used where there may be exposure to temperatures below freezing, but where the brick is not likely to be permeated with water?
    A. Grade MW
    B. Grade SW
    C. Grade NW
    D. None of the above.

13. _____ The type of brick that is used on exposed surfaces where appearance is an important consideration is the _____ brick.
    A. building
    B. facing
    C. firebox
    D. hollow

14. _____ *True or False?* Facing bricks are identical to hollow bricks but have a larger core area.

15. _____ Which of the following types of pavers can be exposed to high levels of pedestrian traffic, such as in stores or exterior walkways?
    A. Type I
    B. Type II
    C. Type III
    D. None of the above.

16. Type _____ heavy vehicular paving brick is intended to be set in a base of sand with sand joints.

_____

17. _____ *True or False?* Ceramic glazed brick and glazed brick, single fired, have identical characteristics and strengths.

18. _____ Fired clay units with normal face dimensions but a reduced thickness are called _____.
    A. brick panels
    B. ceramic glazed brick
    C. glazed brick
    D. thin brick veneer

19. _____ Brick _____ are a combination of thin brick, reinforcement, insulation, and concrete created in a factory setting.
    A. veneers
    B. panels
    C. floorings
    D. walls

20. _____ *True or False?* Firebox bricks must pass a number of assessments, including a rupture test, size test, and warpage test.

21. _____ Chemical-resistant masonry units are made for locations where chemicals and _____ are present at a significant level.
    A. dust
    B. silica
    C. acids
    D. moisture

22. _____ Which of the following brick is intended for use in drainage structures?
    A. Industrial floor brick
    B. Sewer and manhole brick
    C. Chemical-resistant masonry units
    D. All of the above.

Name _____

23. _____ Industrial floor bricks fall into which four categories?
    A. Type C, L, M, and T
    B. Type H, L, M, and T
    C. Type H, K, M, and T
    D. Type H, L, N, and T

24. _____ Brick sizes can be separated into two groups—modular and _____.
    A. permanent
    B. molded
    C. nonmodular
    D. large

25. _____ *True or False?* A modular unit is based on a measurement of 5″.

26. _____ Nominal dimensions of modular bricks are equal to the manufactured dimensions plus the thickness of the _____.
    A. mortar joints
    B. brick
    C. facing
    D. layers

27. _____ ASTM specifications state that texture and color of brick and structural clay facing tile shall conform to an approved sample showing the full range of color and texture. In general, _____ samples are required.
    A. one to three
    B. two to four
    C. three to five
    D. four to six

28. _____ *True or False?* The term *structural clay products* generally refers to burned clay units that are used primarily in building construction.

29. _____ Structural clay products may help support the structure or may serve only as a _____.
    A. backup support
    B. decorative finish
    C. nonloadbearing support
    D. loadbearing support

30. _____ The Federal Trade Commission states that the composition of structural clay products is primarily clay, _____, mixtures.
    A. silica
    B. feldspar
    C. shale
    D. flint

31. _____ A solid masonry unit is one whose cross-sectional area in every plane parallel to the bearing surface is _____ or more of its gross cross-sectional area measured in the same plane.
    A. 50%
    B. 65%
    C. 75%
    D. 80%

32. _____ The tops or bottoms of bricks or blocks are referred to as the _____ surfaces.
    A. bearing
    B. external
    C. plane
    D. outside

33. _____ A brick with less than 75% solid material in its net cross-sectional area in any plane is called a _____ masonry unit.
    A. solid
    B. hollow
    C. structural
    D. None of the above.

34. _____ *True or False?* Soluble salts can affect the weathering of a brick.

35. _____ Efflorescence collect on a wall's surface as the water _____.
    A. condensates
    B. soaks the wall
    C. evaporates
    D. runs down the wall

36. _____ *True or False?* The ability of a masonry unit to stand up under heavy weight without crumbling is called tensile strength.

37. _____ Which of the following statements is *false*?
    A. Most bricks have compressive strengths below 4500 psi.
    B. Typical mortars have compressive strengths from 750 to 2500 psi.
    C. Compressive strength is affected by the strength of the mortar.
    D. Compressive strength is affected by quality of workmanship.

38. The word *bond*, as it refers to masonry, may mean structural bond, mortar bond, or _____ bond.

_____

39. _____ One of the ways structural bonding in brick masonry may be achieved is to arrange the bricks in a(n) _____ fashion.
    A. adjacent
    B. buried
    C. continuous
    D. overlapping

40. _____ What is illustrated in the following image?
    A. Rowlock
    B. Course
    C. Wythe
    D. Header

Goodheart-Willcox Publisher

41. _____ What is illustrated in the following image?
    A. Rowlock
    B. Course
    C. Wythe
    D. Header

Goodheart-Willcox Publisher

Name _____

42. View the following images and identify the correct term for the cut or broken brick pieces shown.

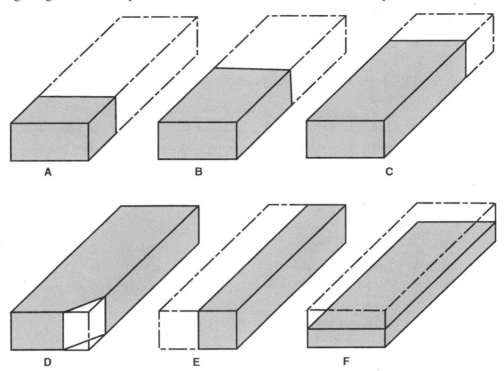

A. _____

B. _____

C. _____

D. _____

E. _____

F. _____

43. Structural bonding of masonry walls with metal ties is acceptable for solid or _____ wall construction.

_____

44. _____ *True or False?* Reinforced brick masonry walls can be structurally bonded by pouring grout into the cavity between wythes of masonry.

45. _____ *True or False?* Grout is usually recommended for bonding of exterior walls.

46. _____ Which bond is shown in the following image?
 A. Flemish bond
 B. English bond
 C. Running bond
 D. Common bond

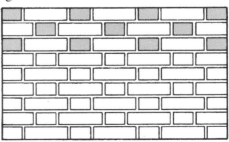

Chapter 7 Clay Masonry Materials 29

47. _____ _____ bond has a course of full-length headers at regular intervals to provide structural bonding as well as pattern.
    A. Running
    B. Dutch
    C. Flemish
    D. Common

48. In a Flemish bond, if the heads are not needed for structural bonding, half bricks may be used. These half bricks are called clipped or _____ headers.

    _____

49. _____ The Flemish bond that has three stretchers alternating with a header is known as a _____ wall bond.
    A. design
    B. garden
    C. Dutch
    D. triple

Match each type of mortar joint with its definition.

50. _____ Created by striking off the excess mortar on a joint, then running a steel jointing tool with a concave surface.

51. _____ Formed by striking a joint, then running a V-shaped steel joint tool to compact the mortar.

52. _____ Made by holding the trowel flat against the brick and cutting in any direction; joint is not compacted.

53. _____ Made by removing some mortar while it is soft; creates a warm shadowing effect.

54. _____ Made by squeezing out the mortar as the unit is laid; mortar is not trimmed off.

55. _____ Requires care since it must be worked from below; it is a compacted joint that readily sheds water.

56. _____ Joint is struck with the trowel, and some compacting occurs.

A. Concave joint
B. Extruded joint (weeping joint)
C. Raked joint
D. Rough cut joint (flush joint)
E. Struck joint
F. V-shaped joint
G. Weather joint

57. _____ *True or False?* Custom brick shapes can be produced in extruded, molded, or handmade brick.

58. _____ *True or False?* Structural clay tiles are produced as loadbearing and nonloadbearing types.

59. _____ When structural clay tile is to be laid with the cells in a horizontal plane, the unit is called _____ construction.
    A. edge
    B. end
    C. finish
    D. side

60. _____ When structural clay tile is to be laid with the cells in a vertical plane, the unit is called _____ construction.
    A. edge
    B. end
    C. finish
    D. side

61. _____ One of the surface textures of structural clay tile is wire cut, which is also known as _____ finish in some regions.
    A. crisscross
    B. engraved
    C. universal
    D. scored

Name _____

62. _____ *True or False?* Glazed clay facing tile is made from shale and other darker-burning clays.

63. _____ Which of the following are the classifications for unglazed facing tile?
   A. FTS and FTX
   B. LB and LBX
   C. MW and NW
   D. S and G

64. The name for the structural facing tile that is designed to absorb sounds is called _____ acoustile.

_____

65. _____ The three basic shapes of structural facing tile include all of the following, *except* _____.
   A. corners and closures
   B. octagons
   C. starters and miters
   D. stretchers

66. _____ Architectural terra cotta is fired clay, usually _____ in color.
   A. brownish-blue
   B. brownish-orange
   C. brownish-red
   D. brownish-yellow

67. _____ *True or False?* Architectural terra cotta is a custom product.

68. _____ _____ ceramic veneer is usually produced in thin sections.
   A. Anchored
   B. Adhesion
   C. Solid
   D. None of the above.

69. _____ During firing of the terra cotta, the clay products are heated to _____ in kilns for three to five days.
   A. 500°F
   B. 110°F
   C. 2200°F
   D. 3000°F

Notes

Name _____ Date _____ Class _____

CHAPTER 8
Concrete Masonry Units

Carefully read Chapter 8 of the text and answer the following questions.

1. _____ Concrete masonry units are hollow or solid units made from a relatively dry mixture of Portland cement, _____, water, and often admixtures.
    A. mud
    B. plastics
    C. aggregates
    D. chemicals

2. _____ The term that refers to sand and gravel or a suitable substitute when referring to concrete masonry units is _____.
    A. accumulate
    B. admixture
    C. aggregate
    D. cinder

3. _____ In concrete masonry units, coloring agents, air-entraining materials, accelerators, retarders, and water repellents are called _____.
    A. aggregates
    B. admixtures
    C. scoria
    D. mortar joints

4. _____ A concrete mixture is molded into desired shapes through compaction and _____.
    A. mixing
    B. melting
    C. molding
    D. vibration

5. _____ Aggregates normally make up about _____ of concrete masonry units by weight.
    A. 60%
    B. 70%
    C. 80%
    D. 90%

6. _____ *True or False?* Aggregates can resist the forces exerted by freezing, thawing, expansion, and contraction.

7. _____ The scrap left after the melting of metals is called _____.
    A. fly ash
    B. slump
    C. scoria
    D. None of the above.

8. _____ Concrete masonry units are classified into _____ main groups based on their intended use, size, and appearance.
    A. four
    B. five
    C. six
    D. seven

9. _____ *True or False?* Masonry units intended primarily for construction of concrete masonry walls, beams, and columns are covered by ASTM standards.

10. _____ *True or False?* In the ASTM standard number ASTM C90-14, the number 14 is the fixed designation for that standard.

11. _____ Concrete bricks are completely solid or have a shallow depression called a(n) _____.
    A. frog
    B. indentation
    C. slide
    D. tray

12. _____ The two grades of concrete bricks included in ASTM C55 are Grade N and Grade _____.
    A. S
    B. C
    C. T
    D. A

13. _____ Slump brick is produced from a mixture that is _____ enough to cause the units to sag when removed from the molds.
    A. cold
    B. dry
    C. hot
    D. wet

14. _____ A solid concrete block is one in which the hollow parts in a cross section are not more than _____ of the total cross-sectional area.
    A. 15%
    B. 20%
    C. 25%
    D. 30%

15. _____ Compressive strength of CMU is based on gross area of the units, with a minimum average of _____ pounds per square inch requirement for a test sample of three blocks.
    A. 200
    B. 500
    C. 2000
    D. 3000

16. _____ *True or False?* Hollow loadbearing blocks are most often used to support great loads.

17. _____ Which of the following combines high compressive strength with light weight and flexible design, size, and shape?
    A. Hollow loadbearing block
    B. Hollow nonloadbearing block
    C. Solid loadbearing block
    D. Solid nonloadbearing block

18. Provisions found in ASTM _____ relating to finish and appearance prohibit defects that impair the strength or permanence of the construction.

_____

19. _____ When referring to concrete building units, it is common practice to identify the size of the unit in which order?
    A. Height, length, width
    B. Length, height, width
    C. Width, height, length
    D. Width, length, height

20. _____ Sizes of concrete building units are usually given in their nominal dimensions. What are the nominal dimensions of a unit measuring 7 5/8″ wide, 7 5/8″ high, and 15 5/8″ long?
    A. 7″ × 7″ × 15″
    B. 8″ × 8″ × 16″
    C. 35/8″ × 35/8″ × 75/8″
    D. None of the above.

21. _____ *True or False?* A two-core design uses lighter block units compared to a three-core design.

Name _____

22. Review the following images and identify concrete blocks shown.

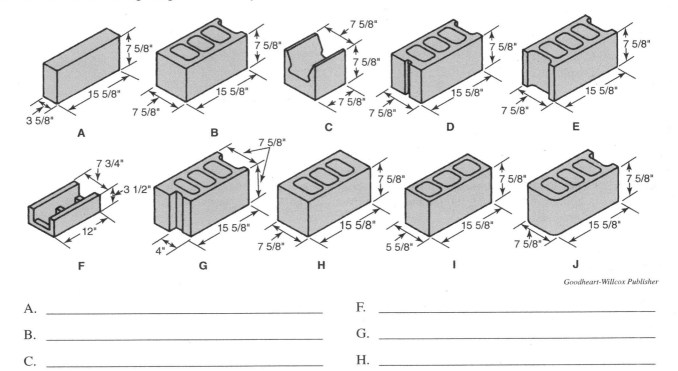

Goodheart-Willcox Publisher

A. _____ F. _____

B. _____ G. _____

C. _____ H. _____

D. _____ I. _____

E. _____ J. _____

23. _____ *True or False?* Concrete blocks may be plain or flat on the end or concave in shape.

24. _____ Concrete blocks with concave ends have two _____.
    A. caps
    B. ears
    C. points
    D. cells

25. _____ The openings in concrete blocks are referred to as _____.
    A. cells
    B. cores
    C. voids
    D. All of the above.

26. _____ The wall between the cells on a concrete block is called the _____.
    A. header
    B. cross web
    C. face web
    D. face shell

27. The outside face of a block is called the face _____.

_____

28. _____ *True or False?* Coarse surface textures on a block absorb more sound and take less paint than smooth textures.

29. _____ Two components that affect the color of concrete masonry units are the color of the ingredients used and the _____ method.
    A. mixing
    B. curing
    C. heating
    D. cooling

30. _____ CMUs that have specially prepared surface material applied to one or more surfaces to provide color, pattern, or texture are called _____ concrete masonry units.
   A. prefaced
   B. ground face
   C. architectural
   D. split face

31. _____ The applicable standard for prefaced concrete masonry units is ASTM _____.
   A. C91
   B. C150
   C. C744
   D. C876

32. _____ Concrete masonry units that include oversized and specialty units that may be used to create walls, columns, and arches of monumental proportions are called _____ units.
   A. prefaced
   B. ground face
   C. architectural
   D. split face

33. _____ Masonry units that have been designed for aesthetic appeal are called _____ concrete masonry units.
   A. architectural
   B. decorative
   C. ornamental
   D. striking

34. _____ *True or False?* Split face masonry units look like rough quarried stone and are produced by splitting solid concrete lengthwise.

35. _____ What is the name for the type of concrete block that helps control light and privacy?
   A. Barrier block
   B. Divider block
   C. Partition block
   D. Screen block

36. _____ _____ materials are made similarly to concrete block but are usually made of solid materials to survive severe weather.
   A. Hardscape
   B. Insulated
   C. Freestanding wall blocking
   D. None of the above.

37. What type of concrete block is shown in the following image?

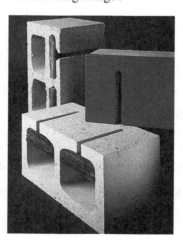

Trenwyth Industries, Inc.

Name _____

38. Insulated blocks are concrete blocks that have plastic foam inserts that increase their resistance to _____ gain and loss.

_____

39. Review the following images and identify the pattern bonds shown.

Goodheart-Willcox Publisher

A. _____

B. _____

C. _____

D. _____

E. _____

F. _____

40. _____ Sand-lime bricks have uniform size, shape, and _____.
    A. texture
    B. weight
    C. color
    D. spacing

41. _____ Grade _____ sand-lime brick can be exposed to freezing temperatures and water saturation.
    A. NW
    B. MW
    C. SW
    D. S

42. _____ The designs on glass block distribute and control the direction of _____.
    A. airflow
    B. dust
    C. light rays
    D. noise

43. _____ What are the three categories of glass blocks?
    A. Decorative, functional, general purpose
    B. Depressed, offset, sculptured
    C. Enhanced, fluted, ribbed
    D. Standard, enhanced, special purpose

44. _____ Glass blocks are manufactured in these three popular sizes, _____ inches square.
    A. 4 3/4, 6 3/4, and 11 3/4
    B. 4 3/4, 7 3/4, and 10 3/4
    C. 5 3/4, 7 3/4, and 11 3/4
    D. 6 3/4, 7 3/4, and 12 3/4

45. _____ *True or False?* Soft conversion means inch-pound dimensions are stated in metric units, but the material sizes currently produced are not required to change.

Name _____ Date _____ Class _____

CHAPTER 9
Stone

Carefully read Chapter 9 of the text and answer the following questions.

1. _____ *True or False?* In construction today, stone is almost always used as a structural material.

2. _____ Which of the following is *not* one of the classifications of stone?
   A. Igneous
   B. Sedimentary
   C. Metamorphic
   D. None of the above.

3. _____ When igneous stone is composed primarily of light materials (quartz and feldspar), it is called _____.
   A. acidic
   B. basic
   C. caustic
   D. magma

4. _____ Which of the following is an igneous stone that is formed mainly of potash, feldspar, quartz, and mica?
   A. Granite
   B. Limestone
   C. Sandstone
   D. Traprock

5. _____ *True or False?* Traprock is the quarryman's term for diabase, basalt, and gabbro.

6. _____ Stone that is formed in layers with deposits of shell, disintegrated stone, or sand becoming cemented together under pressure is called _____ stone.
   A. igneous
   B. metamorphic
   C. sedimentary
   D. sand

7. _____ Which of the following is a stone that is composed mainly of grains of quartz cemented by silica, lime, or iron oxide?
   A. Granite
   B. Limestone
   C. Sandstone
   D. Slate

8. _____ _____ cement can produce a hard, durable sandstone.
   A. Portland
   B. Silica
   C. Porous
   D. Colored

9. _____ A sedimentary stone that consists mainly of the calcium carbonate material calcite and is usually marine in origin is _____.
   A. limestone
   B. marble
   C. sandstone
   D. traprock

10. _____ A high-quality, widely used building stone is limestone from south-central Indiana, called Indiana limestone or _____ limestone.
    A. white
    B. Amherst
    C. Bedford
    D. natural bed

11. _____ *True or False?* When limestone is set with the grain running vertically, it is said to be set on its natural bed.

12. _____ The stone that is formed through reconstitution due to great heat and pressure is the _____ rock.
    A. igneous
    B. sedimentary
    C. trap
    D. metamorphic

13. _____ Which type of metamorphic stone is recrystallized limestone that can be white, yellow, brown, green, or black?
    A. Gneiss
    B. Marble
    C. Quartzite
    D. Schist

14. _____ The metamorphic stone that can be split into sheets and used for roofing or flagstones is called _____.
    A. marble
    B. slate
    C. gneiss
    D. quartzite

15. _____ The rather coarse-grained stone that contains large amounts of mica is _____.
    A. gneiss
    B. marble
    C. slate
    D. schist

16. _____ _____ is a stone that is difficult to define or describe because it is so varied.
    A. Gneiss
    B. Schist
    C. Quartzite
    D. Marble

17. _____ Sandstone that has been recrystallized is called _____.
    A. gneiss
    B. schist
    C. quartzite
    D. marble

18. _____ What is the lowest quality of the grades of stone?
    A. Rustic
    B. Statuary
    C. Standard
    D. Select

19. There are two color descriptions of Indiana limestone—buff and _____.
    _____

20. _____ The Indiana Limestone Institute of America (ILI) classifies Indiana limestone into _____ colors and _____ grades based on granular texture and other natural characteristics.
    A. two; four
    B. three; five
    C. four; two
    D. five; three

Name _____

21. _____ Which surface finish results in a moderately smooth surface that is inexpensive because it requires no further finishing after leaving the saw?
    A. Gang sawed
    B. Machine tooled
    C. Plucked
    D. Shot sawed

22. _____ The surface finish that has two to ten grooves per inch that are parallel and concave in shape and may be used on any type of stone is called _____.
    A. gang sawed
    B. machine tooled
    C. shot sawed
    D. hand tooled

23. _____ When special accents are required, which of the following types of surface finishes should be used?
    A. Gang sawed
    B. Hand tooled
    C. Plucked
    D. Rubbed and honed

24. _____ Which surface finish should be used to obtain a smooth finish and is used primarily for limestone?
    A. Carborundum
    B. Gang sawed
    C. Plucked
    D. Shot sawed

25. _____ *True or False?* Uncoursed stone is stone laid in an irregular pattern in which stones are placed randomly in the wall.

26. _____ Which of the following is a stone that consists of broken fragments of irregular shapes and textures?
    A. Ashlar
    B. Dimensioned stone
    C. Roughly squared stone
    D. Rubble

27. _____ *True or False?* In a roughly squared stone uncoursed pattern, both large and small stones may be used.

28. _____ Stone cut to specific dimensions, dressed and finished to precise job requirements at the mill, and then transported to the site as a finished product is called _____.
    A. ashlar
    B. machined
    C. roughly squared stone
    D. rubble

29. _____ _____ is an excellent sill material that can be used on exterior masonry walls.
    A. Sandstone
    B. Limestone
    C. Slate
    D. Marble

30. _____ Which of the following is the name for large squared stones set at the corner of buildings?
    A. Copings
    B. Pavers
    C. Quoins
    D. Trim

31. _____ If a structure has outside corners, specially manufactured _____ stones are employed to cover the corners.
    A. coping
    B. corner
    C. roughly squared
    D. slate

32. _____ Manufactured stone cannot be applied directly to a base coat of any masonry surface without the use of wire mesh or _____.
    A. metal lath
    B. WRBs
    C. masonry adhesive
    D. mortar

33. _____ Mortar used for manufactured stone should be a mixture of _____ part cement to _____ parts clean sand.
    A. one; two
    B. one; three
    C. one; four
    D. one; five

34. After the mortar joints on a wall become firm, the joints should be _____ up.
_____

Name _____ Date _____ Class _____

CHAPTER 10
Mortar and Grout

Carefully read Chapter 10 of the text and answer the following questions.

1. _____ Which of the following is the term for the bonding agent that ties masonry units into a strong, well-knit, weathertight structure?
   A. Mortar
   B. Glue
   C. Grout
   D. Cement

2. _____ Basic mortar includes what three materials?
   A. Portland cement, hydrated lime, and sand.
   B. Portland cement, burned gypsum, and sand.
   C. Masonry cement, hydrated lime, and grout.
   D. Masonry cement, quicklime, and sand.

3. _____ *True or False?* One function of mortar is to make up for the difference in sizes of units.

4. _____ Material properties of mortar that influence the structural performance of masonry are listed below. Which of the following is the *least* important?
   A. Bond strength
   B. Compressive strength
   C. Elasticity
   D. Workability

5. The standards for materials commonly used in mortars are provided by _____ International.

   _____

6. _____ Portland cement is a hydraulic material, meaning that it _____.
   A. cannot be mixed with water alone
   B. comes premixed with water
   C. hardens under water
   D. softens under water

7. _____ *True or False?* Different brands of masonry cement are manufactured with similar ingredients and can be changed during the construction of a specific project.

8. Hydrated lime is quicklime that has been _____ before packaging.

   _____

9. _____ Type S hydrated lime is recommended for masonry mortar because unhydrated oxides and _____ are controlled.
   A. hardness
   B. compressive strength
   C. plasticity
   D. All of the above.

10. _____ The primary aggregate used in masonry mortar is _____.
    A. stone
    B. sand
    C. mortar
    D. clay

11. _____ The separation of water during the time that mortar is in the plastic state prior to setting is called _____.
    A. division
    B. bleeding
    C. hydration
    D. None of the above.

12. _____ Water for masonry is required by ASTM C270 to be clean and free of harmful amounts of acids, _____, or organic materials.
    A. alkalis
    B. sand
    C. chemicals
    D. cement

13. _____ *True or False?* Water from city mains or private wells is generally suitable for mixing mortar.

14. _____ *True or False?* When mixing mortar, consistency with respect to yield, workability, and color is important.

15. _____ According to ASTM C270, 1 cu ft of loose, damp sand contains _____ lb of dry sand.
    A. 50
    B. 60
    C. 70
    D. 80

16. _____ Most sand on a construction site contains from _____ moisture content.
    A. 3% to 7%
    B. 4% to 8%
    C. 5% to 9%
    D. 6% to 10%

17. Cubic foot measuring _____ are recommended for measurement of sand on the jobsite.

_____

18. _____ *True or False?* When possible, mortar should be mixed by hand.

19. _____ Ingredients should be mixed in a mortar mixer for _____ minutes.
    A. 2 to 4
    B. 3 to 5
    C. 4 to 6
    D. 5 to 7

20. _____ _____ mortar boxes do not rust and are less likely to have mortar stick to the box.
    A. Steel
    B. Polyethylene
    C. Sand
    D. All of the above.

21. _____ Workability of mortar can be restored by adding lost water and mixing briefly. This is called _____.
    A. dampening
    B. hydrating
    C. retempering
    D. bleeding

22. _____ *True or False?* Plastic mortar is uniform, is cohesive, and has a usable consistency.

23. _____ Water retentivity of plastic mortar improves with the addition of fine sand and with higher _____ content.
    A. moisture
    B. lime
    C. chemical
    D. aggregate

Name _____

24. View the following image of slump test comparisons. Identify the desired slump for mortar, grout, and concrete.

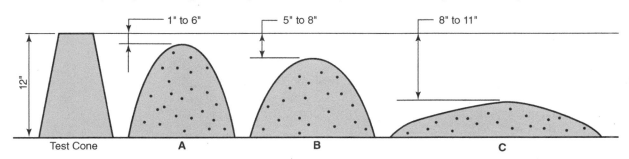

A. _____

B. _____

C. _____

25. _____ The compressive strength of mortar _____ as cement content is increased and _____ as lime is increased.
    A. decreases; decreases
    B. decreases; increases
    C. increases; decreases
    D. increases; increases

26. The property of hardened mortar that holds the masonry units together is called _____.

_____

Match the mortar type with its correct definition.

27. _____ High strength; used for unreinforced masonry below grade.

28. _____ Low strength; general interior use in nonloadbearing masonry; not recommended by the MSJC Code.

29. _____ Medium-high strength; used where bond and lateral strength are more important than compressive strength.

30. _____ Medium-strength; suitable for use in exposed masonry above ground.

31. _____ Very low strength; used for interior nonloadbearing partition walls where strength is not needed; not recommended by the MSJC Code.

A. Type M
B. Type S
C. Type N
D. Type O
E. Type K

32. _____ _____ mortar is beneficial for training because it can be reused again and again.
    A. Type N
    B. Type O
    C. Educational
    D. Type M

33. _____ *True or False?* Antifreeze compounds and water-repellent agents should be added to mortar as needed.

34. _____ For cavity walls in areas where wind velocities may exceed 80 mph, _____ mortar should be used.
    A. Type M
    B. Type S
    C. Type N
    D. Type O

35. _____ For facing tile, any mortar type *except* _____ is acceptable.
    A. Type S
    B. Type N
    C. Type O
    D. Type K

36. _____ The process of filling a mortar joint with fresh mortar is called _____.
    A. tuckpointing
    B. facing
    C. pointing
    D. bonding

37. _____ The complete process of repairing a defective masonry joint is called _____.
    A. facing
    B. repointing
    C. bonding
    D. tuckpointing

38. _____ The term for a coloring agent that can be added to replacement mortar to make it match the original mortar is _____.
    A. dye
    B. pigment
    C. hue
    D. shade

39. _____ Dirt-resistant mortar has aluminum tristearate, calcium stearate, or ammonium stearate added. The amount added should be _____ of the weight of the Portland cement.
    A. 1%
    B. 3%
    C. 5%
    D. 7%

40. _____ *True or False?* Grout is used in brick masonry to fill cells of hollow units or spaces between wythes of solid units.

41. _____ Grouts with _____ initial water content exhibit more shrinkage than grouts with _____ water content.
    A. low; high
    B. high; low
    C. clean; dirty
    D. dirty; clean

42. _____ The ideal temperature of mortar is between _____.
    A. 40°F and 60°F
    B. 50°F and 70°F
    C. 60°F and 80°F
    D. 70°F and 90°F

43. _____ Water or sand should not be heated above 140°F (60°C) to prevent the danger of _____, which occurs when the heated water and sand come in contact with the cement that is part of the mix.
    A. explosion
    B. flash set
    C. boiling
    D. fusion

44. _____ Glass units should *not* be laid when temperatures are below _____.
    A. 32°F
    B. 40°F
    C. 90°F
    D. 115°F

Name _____

45. _____ The process of mortar or grout losing moisture to extremely hot and dry masonry units is called _____.
    A. bleeding
    B. slaking
    C. dryout
    D. plasticity

46. _____ When mortar quantities are estimated, it is important to remember that _____ of cementitious material and _____ of sand make 1 cu ft of mortar.
    A. 0.33 cu ft; 0.66 cu ft
    B. 0.33 cu ft; 0.99 cu ft
    C. 0.66 cu ft; 0.33 cu ft
    D. 0.99 cu ft; 0.33 cu ft

47. _____ *True or False?* The only reason to order more mortar than calculations indicate is if change orders require more materials.

48. _____ Estimating mortar for a typical single-wythe brick wall can be more confusing than estimating for a concrete block wall because more _____ exist and mortar coverage is different.
    A. sizes
    B. shapes
    C. textures
    D. All of the above.

49. _____ *True or False?* Grout bonds steel and masonry units together so they function as a single unit.

50. _____ Grouting brick and block walls _____.
    A. bonds the wythes of masonry together
    B. increases the cross-sectional area of the wall
    C. transfers stress from the masonry to the reinforcing steel when lateral forces are present
    D. All of the above.

51. _____ Grout used in masonry wall construction should comply with the requirements of ASTM _____.
    A. C5
    B. C150
    C. C404
    D. C476

52. _____ When concrete masonry units are highly absorbent, use of a grouting-aid _____ can reduce early water loss to the masonry units.
    A. aggregate
    B. admixture
    C. absorbent
    D. cement

53. _____ Grout should be fluid enough to be pumped or poured, but not so fluid as to cause _____.
    A. dripping
    B. soaking
    C. segregation
    D. sticking

54. Grout strength should not be less than _____ psi.

_____

55. _____ *True or False?* Water that is absorbed by the masonry units before the grout has set increases the water-cement ratio and decreases the compressive strength.

56. _____ *True or False?* Grout should be placed within 1 1/2 hours after water is first added.

57. _____ Grout is consolidated by vibration or _____.
    A. segregation
    B. shrinkage
    C. pouring more
    D. rodding

58. _____ The process of controlling the moisture loss of cementitious material over time to attain designed standards of strength and bond adherence is called _____.
    A. low-lift grouting
    B. high-lift grouting
    C. curing
    D. sampling

59. For sampling and testing, ASTM Standard _____ should be used for quality control and as a guide for selecting grout proportions.

_____

Name _____ Date _____ Class _____

CHAPTER 11
Anchors, Ties, and Reinforcement

Carefully read Chapter 11 of the text and answer the following questions.

1. _____ *True or False?* Masonry anchors, ties, and joint reinforcements are placed inside a masonry wall to give it greater strength or to hold it in place.

2. _____ Strips of metal or metal wire used to tie masonry wythes together or to tie masonry veneer to a concrete frame or wood frame wall are called _____ ties.
    A. frame
    B. masonry
    C. stone
    D. wall

3. _____ Which of the following is *not* one of the three types of unit ties that are used in masonry wall construction?
    A. Corrugated ties
    B. Rectangular ties
    C. Square ties
    D. Z ties

4. _____ Continuous horizontal joint reinforcement is available in lengths of _____.
    A. 6' to 8'
    B. 10' to 12'
    C. 14' to 16'
    D. 12' to 24'

5. _____ Which style of horizontal joint reinforcement is shown in the following illustration?
    A. Truss
    B. Tab
    C. Ladder
    D. Adjustable

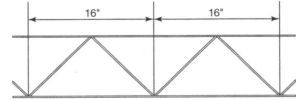

Goodheart-Willcox Publisher

6. _____ Ties that are used to attach a masonry facing to a backup system of masonry or typical stud construction are called _____ ties.
    A. adjustable
    B. flexible
    C. surface
    D. supply

7. _____ *True or False?* Adjustable ties generally provide more strength and stiffness than unit ties or joint reinforcement.

8. _____ Typical re-anchoring approaches include mechanical expansion systems, screw systems, and _____ systems.
    A. automatic extension
    B. epoxy adhesive
    C. resin
    D. nail

Copyright Goodheart-Willcox Co., Inc.
May not be reproduced or posted to a publicly accessible website.

49

9. _____ Reinforcing rods (re-rods) or bars (rebars) are reinforcing steel used to increase the _____ strength of concrete.
    A. tensile
    B. compressive
    C. performance
    D. All of the above.

10. _____ *True or False?* When reinforcing is used in a structure, steel and masonry act together to resist forces being exerted on the wall or surface.

11. _____ Reinforcing bars can be placed in masonry walls in which of the following orientations?
    A. Diagonally only.
    B. Horizontally only.
    C. Vertically or diagonally.
    D. Vertically or horizontally.

12. _____ Common practice is to place vertically reinforcing bars every _____′ in footings to tie the footings to the concrete block wall.
    A. 2
    B. 4
    C. 6
    D. 20

13. _____ The sizes of reinforcing bars permitted by the MSJC Code are indicated by _____.
    A. letters
    B. numbers
    C. signs
    D. symbols

14. _____ Reinforcing bars have _____ on the surface that aid in bonding.
    A. coatings
    B. identification marks
    C. deformations
    D. adhesive

15. _____ Grade _____ and grade _____ bars show the identification marks of the producing mill, bar size number, and type.
    A. 40; 50
    B. 60; 75
    C. 40; 60
    D. 50; 75

16. _____ *True or False?* Anchors usually connect two similar materials.

17. _____ One type of anchor is corrugated with a _____ on one end for veneer construction.
    A. bolt
    B. band
    C. dovetail
    D. strap

18. _____ Which of the following devices are placed in fresh concrete and used to attach sill plates to masonry work?
    A. Anchor bolts
    B. Corrugated anchors
    C. Reinforcing bars
    D. Strap anchors

19. _____ Bearing walls that intersect are frequently connected with a _____ anchor.
    A. corrugated
    B. strap
    C. hex coupling
    D. flat head

Name _____

20. _____ *True or False?* Reinforcing bars are intended for installation after the concrete has hardened and generally involve some type of expansion device.

21. _____ Joint reinforcement is especially designed for placement in the horizontal joints of masonry walls to provide greater _____ strength.
    A. tensile
    B. compressive
    C. lateral
    D. diagonal

22. _____ The maximum size of longitudinal and cross wires is _____ the mortar joint thickness.
    A. one-fourth
    B. one-half
    C. two-thirds
    D. three-quarters

23. _____ Which of the following is another name for hardware cloth?
    A. Lattice wire cloth
    B. Mesh wall tie
    C. Net wall tie
    D. Web wire cloth

Notes

Name _____ Date _____ Class _____

CHAPTER 12
Laying Brick

Carefully read Chapter 12 of the text and answer the following questions.

1. _____ The _____ determines quantities needed for a construction project and orders the materials.
   A. supervisor
   B. estimator
   C. contractor
   D. mason

2. _____ *True or False?* One of the first brick operations to occur is organizing the site for brick laying.

3. _____ *True or False?* Placing a brick unit is easier if an excess amount of mortar has been laid down.

4. _____ Which of the following is *not* a method used to load mortar on a trowel?
   A. From the side of the board.
   B. From the top of the pile.
   C. From the middle.
   D. From under the pile.

5. _____ To hold a trowel correctly, grasp it in your dominant hand with fingers under the handle and the thumb on top of the _____.
   A. ferrule
   B. cap
   C. shank
   D. heel

6. _____ In the following image, the mason is _____ the first course to form a uniform bed.
   A. cross jointing
   B. furrowing
   C. tempering
   D. buttering

Author's image taken at Job Corps, Denison, IA

7. Another name for the brick head joint in courses being laid is _____ joint.

_____

8. _____ Mortar that is not used within _____ hour(s) should be discarded.
   A. 1
   B. 1 1/2
   C. 2
   D. 2 1/2

9. _____ *True or False?* When laying brick on the mortar bed, it is important to hold the line against the side of the brick being placed.

10. _____ Using both hands at the same time when laying brick increases efficiency and reduces _____.
    A. fatigue
    B. brick waste
    C. mortar waste
    D. brick damage

11. _____ *True or False?* Mortar for a head joint should be applied to the brick after it is placed on the mortar bed.

12. _____ The application of mortar on a masonry unit with a trowel is referred to as _____.
    A. buttering
    B. furrowing
    C. plaster back
    D. smearing

13. _____ Which of the following tools for cutting brick produces a rough cut but is acceptable for applications where the cut edge is hidden by mortar?
    A. Brick hammer
    B. Brick set chisel
    C. Masonry saw
    D. Trowel

14. _____ A _____ can be used to cut soft brick but is not recommended for hard brick.
    A. brick chisel
    B. brick hammer
    C. trowel
    D. masonry saw

15. _____ When an exact, smooth cut is required, brick is cut with a _____.
    A. brick chisel
    B. brick hammer
    C. trowel
    D. masonry saw

16. _____ *True or False?* When checking bricks with a level, you should tap on the level if the bricks are not level side to side.

17. _____ *True or False?* A wall is usually laid to a line if the wall is longer than the plumb rule.

18. _____ When a wall is being built, a _____ is usually built first at each corner to establish proper height and provide a place to attach a mason's line.
    A. joint
    B. lead
    C. course
    D. corner block

19. _____ How far away from the line should brick be laid?
    A. A brick's width away
    B. A brick's height away
    C. A line's width away
    D. Brick can be laid on the line.

20. _____ Which of the following is the correct term for the boards and stakes that locate and preserve the building lines during excavation and construction?
    A. Drop lines
    B. Batter boards
    C. Gauge sticks
    D. None of the above.

Name _____

21. _____ The 3-4-5 method is used to ensure that the layout of the first corner is _____.
    A. plumb
    B. square
    C. level
    D. tight

22. _____ A _____ is a mason's line attached to a plumb bob to transfer line intersections down into the excavation to locate footings and walls.
    A. builder's transit
    B. batter board
    C. drop line
    D. chalk box

23. _____ The pole used to measure the mortar joint heights of each course of masonry is called a _____.
    A. corner pole
    B. mason's pole
    C. story pole
    D. Both A and C.

24. _____ *True or False?* A gauge stick is similar to a corner pole, but is larger and more bulky.

25. _____ "Laying the corners" is also called "laying the _____."
    A. angles
    B. bases
    C. courses
    D. leads

26. _____ *True or False?* In a rack back lead, each course is stepped back from the previous course.

27. _____ Spotting the brick helps to reduce _____.
    A. overlapping joints
    B. racking
    C. line sagging
    D. None of the above.

28. _____ It is important to lay out a dry course of brick from corner to corner to eliminate as much _____ as possible.
    A. mortar use
    B. cutting of units
    C. drying time
    D. leveling

29. _____ When building a corner, the sequence to be followed is _____.
    A. level the unit, then line up, then plumb
    B. plumb, then level the unit, then line up
    C. level the unit, then plumb, then line up
    D. line up, then level the unit, then plumb

30. _____ After laying the second course of bricks, check for proper _____ just after the course has been laid but before the brick has been leveled.
    A. plumb
    B. height
    C. jointing
    D. mortar

31. _____ _____ is the bowing out of brick.
    A. Wind
    B. Cave-in
    C. Racking
    D. Belly

32. _____ When striking a wall, use the _____ jointer for the bed joints.
    A. long
    B. short
    C. sled-runner
    D. Both A and C.

33. _____ A common (American) bond consists of courses of _____ between every five to seven courses of _____.
    A. headers; soldiers
    B. headers; stretchers
    C. stretchers; headers
    D. stretchers; rowlock

34. _____ The term _____ refers to the distance up a masonry wall to the top of the brick course and directly under the header course.
    A. header high
    B. stretcher high
    C. plumb
    D. None of the above.

35. _____ What type of wall ties are used to connect intersecting brick walls?
    A. Rectangular ties
    B. Z ties
    C. Corrugated
    D. All of the above.

36. _____ When laying 8″, two-wythe intersecting brick walls, the metal ties are laid after the second course and again on the _____ course in the wall.
    A. third
    B. fourth
    C. eighth
    D. last

37. _____ Which of the following statements about a 12″ solid wall is *false*?
    A. It is essentially an 8″ solid wall with a third wythe added.
    B. The first course is repeated every sixth or seventh course.
    C. This type of wall is rare today but may be used for thermal storage units.
    D. The third course is composed of all headers.

38. _____ The popular Flemish bond is easy to lay and produces a visually interesting wall consisting of _____.
    A. alternate courses of stretchers and headers
    B. alternate headers and stretchers in each course
    C. one course of headers followed by two courses of stretchers
    D. one course of stretchers followed by two courses of headers

39. _____ *True or False?* English bond consists of alternate courses of stretchers and headers.

40. _____ *True or False?* A cavity wall consists of two separate wythes with a 1″ cavity space.

41. _____ Leaky walls result from partially filled mortar joints that can contribute to _____, which is the deterioration or flaking off the brick.
    A. erosion
    B. efflorescence
    C. spalling
    D. cracking

42. _____ Excess mortar that protrudes from the joint on the inside of a wall cavity is called a mortar _____.
    A. fin
    B. bed
    C. dropping
    D. spall

Name _____

43. _____ The process of flattening mortar against the back side of masonry units is called _____.
   A. spalling
   B. furrowing
   C. plaster back
   D. weep hole

44. _____ Weep holes can be created by leaving out one of the _____ joints during construction.
   A. corner
   B. head
   C. spacing
   D. concave

45. The two types of joints that are the most weathertight are _____ joints and concave joints.
_____

46. _____ *True or False?* Truss type joint reinforcement should not be used to tie wythes of a brick and block cavity wall together.

47. _____ Which of the following statements about the storage of materials on the construction site is *false*?
   A. Sand for mortar should be placed off the ground.
   B. Flashing materials should be stored where they will not be punctured or damaged.
   C. Masonry units should be stored on the ground.
   D. Rigid insulation should be stored where it is not exposed to sunlight.

48. A water-resistant membrane should be placed over walls with an overhang of at least _____" on each side.
_____

49. _____ In single wythe brick loadbearing walls, determining the _____ loads is generally the first step in the design of a structural system.
   A. structural
   B. design
   C. architectural
   D. total

50. For single wythe brick bearing wall systems, the MSJC Code limits the maximum size of reinforcement used in masonry to a No. _____ reinforcing bar.
_____

51. _____ In a single wythe brick bearing wall, rigid board insulation is generally placed on the _____ of the brick wythe.
   A. interior face
   B. exterior face
   C. weep holes
   D. top

52. _____ All of the following can be used to limit air leakage in single wythe exterior brick walls, *except* _____.
   A. films
   B. gypsum board
   C. house wraps
   D. paper

53. _____ Which of the following can help prevent water penetration in a single wythe wall?
   A. Full mortar joints with no voids.
   B. A bituminous, dampproof coating on the inside face of the brick wall.
   C. A drainage cavity with flashing and weep holes.
   D. All of the above.

54. _____ The method used to widen a wall by projecting out masonry units to form a ledge or shelf is called _____.
   A. ledging
   B. bedding
   C. corbeling
   D. structuring

55. _____ *True or False?* Corbels normally support a load.

56. _____ When a corbel is being built, the maximum projection of one unit must not exceed _____ the height of the unit or _____ the thickness of the unit at right angles to the wall.
   A. one-half; one-third
   B. one-half; two-thirds
   C. one-third; one-half
   D. two-thirds; one-half

57. _____ Which of the following statements about hollow brick piers is *false*?
   A. They are commonly used as gate posts at corners.
   B. They are shorter than columns.
   C. They are typically constructed using a staggered bond pattern.
   D. They generally support a load.

58. _____ Before constructing a hollow brick pier, a footing and foundation need to be built. The depth of the footing is determined by local or state _____ codes.
   A. digging
   B. excavation
   C. frost
   D. foundation

59. _____ Prior to using any cleaning agent on a masonry wall, apply it to a sample section of approximately _____.
   A. 5 sq ft to 15 sq ft
   B. 10 sq ft to 20 sq ft
   C. 15 sq ft to 25 sq ft
   D. 20 sq ft to 30 sq ft

60. _____ A solution of _____ acid is used extensively for cleaning new masonry.
   A. hydrochloric
   B. cleaning
   C. muriatic
   D. Both A and C.

61. _____ Light-colored brick is _____ to be burned by acid than darker-colored brick.
   A. equally likely
   B. less likely
   C. more likely
   D. None of the above.

62. _____ When applying a solution using a pressurized power washer, lower the pressure to _____ pounds per square foot.
   A. 10 to 30
   B. 20 to 40
   C. 30 to 50
   D. 50 to 60

63. Why does the BIA recommend that abrasive blasting *not* be used for brickwork?

_____

CHAPTER 13
Laying Block

Carefully read Chapter 13 of the text and answer the following questions.

1. _____ *True or False?* Most concrete masonry units (CMUs) are used in the construction of roofing or piers.

2. _____ _____ injuries are one of the leading health issues in the construction industry.
   A. Back
   B. Neck
   C. Shoulder
   D. Leg

3. _____ _____ is the process of preparing the area where the building will be constructed.
   A. Scheduling
   B. Planning
   C. Site preparation
   D. Implementation

4. _____ A _____ is usually one of the first pieces of equipment brought to a worksite to aid management in preparing the job.
   A. scaffold
   B. construction office trailer
   C. bag of premixed mortar
   D. forklift

5. _____ *True or False?* The types of walls that can be built with concrete masonry units are essentially the same as those that are built with clay masonry units.

6. _____ In solid masonry walls, masonry headers, _____, and grout are used between wythes to provide a solid structural bond.
   A. weep holes
   B. mortar
   C. metal ties
   D. stretchers

7. _____ Multiple-wythe hollow masonry walls may be classified as _____ walls.
   A. cavity
   B. composite
   C. veneered
   D. None of the above.

8. _____ A two-wythe wall that allows each wythe to react independently to stress is called a _____ wall.
   A. cavity
   B. composite
   C. hollow masonry
   D. veneered

9. _____ Cavity walls usually consist of two walls separated by continuous air space that is at least _____″ wide.
   A. 1
   B. 2
   C. 4
   D. 6

10. _____ *True or False?* The insulation placed in a cavity wall must be water absorbent.

11. _____ A composite wall is a multi-wythe wall that is designed to act as a single member when subjected to _____.
    A. stress
    B. loads
    C. insulation
    D. Both A and B.

12. _____ Which of the following statements about composite walls is *true*?
    A. Regular ties should be spaced no more than 24″ apart horizontally.
    B. A composite wall is a single-wythe wall.
    C. Distance between headers should not exceed 24″ in hollow masonry.
    D. Adjustable ties should not be spaced more than 16″ apart.

13. _____ *True or False?* Masonry veneer is commonly used as a loadbearing facing material in residential and light commercial construction.

14. _____ *True or False?* An advantage of veneer is its resistance to the effects of wind compared to other wall finishes.

15. _____ Which type of concrete masonry wall is designed for applications that experience high stress, high wind loads, or severe earthquakes?
    A. Cavity
    B. Composite
    C. Reinforced
    D. Veneered

16. _____ Steel embedded in a reinforced concrete masonry wall to increase the strength of the wall is positioned _____.
    A. diagonally
    B. horizontally
    C. vertically
    D. both horizontally and vertically

17. _____ *True or False?* Grouted masonry walls include reinforcements.

18. _____ *True or False?* When grasping a trowel to spread mortar, fingers should be under the handle and the thumb on top of the ferrule.

19. _____ When laying the first course of concrete block, the mortar should be spread and _____ to help form a uniform bed on which to lay the masonry unit.
    A. buttered
    B. furrowed
    C. swiped
    D. cut

20. _____ Mortar is generally bedded only on the outside edges of concrete block. This is called _____ bedding.
    A. face covering
    B. face shell
    C. solid
    D. surface shell

21. _____ *True or False?* If the mortar consistency is too stiff (dry), it will not support the weight of the block.

22. _____ Mortar on the board should be well _____ (sprinkled lightly with water) until it is used.
    A. buttered
    B. furrowed
    C. tempered
    D. bedded

23. _____ Discard mortar that has not been used within _____ hour(s).
    A. 1
    B. 1 1/2
    C. 2
    D. 2 1/2

24. _____ *True or False?* Concrete blocks generally require both hands for placement.

Name _____

25. _____ When making head joints on concrete blocks, form full head joints on both _____ of each block to be laid.
    A. ears
    B. faces
    C. lengths
    D. headers

26. _____ When using a chisel to cut block, _____.
    A. score the block on only one side to obtain a cleaner break
    B. hold the beveled edge away from you
    C. face the piece of block to be cut off away from you
    D. All of the above.

27. _____ To achieve a neat, clean cut when cutting block, a _____ should be used.
    A. blocking chisel
    B. brick hammer
    C. masonry saw
    D. 10″ trowel

28. _____ A mason's line is usually used on any straight wall longer than about _____.
    A. 2′
    B. 4′
    C. 8′
    D. 12′

29. _____ In order to minimize sagging in the mason's line on long walls, a(n) _____ can be set at approximately midpoint.
    A. lead
    B. metal twig
    C. corner pole
    D. anchor

30. _____ *True or False?* A well-planned concrete block structure mainly involves stretcher and corner blocks.

31. _____ What is the first step in laying an 8″ running bond concrete block wall?
    A. Establish the outside wall line.
    B. Lay corner block.
    C. Spread and furrow a full mortar bed.
    D. String out blocks for the first course.

32. _____ *True or False?* It is necessary to align, level, and plumb the blocks only after the full first course has been laid.

33. _____ Which of the following statements is *false*?
    A. Joints can be tooled after the mortar has become thumbprint hard.
    B. The jointing tool should be slightly larger than the width of the joint.
    C. A 1/2″-diameter bar is used for a 3/8″ concave mortar joint.
    D. Long horizontal joints are tooled with a sled runner.

34. _____ Continuous vertical joints in concrete masonry walls where forces might cause cracks are _____ joints.
    A. V
    B. control
    C. header
    D. collar

35. An alternate method of constructing control joints is to place a _____-bar in the horizontal mortar joint above the jamb blocks.

_____

36. _____ What is another name for a control joint that uses building paper to break the bond?
    A. Felt control joint
    B. Michigan control joint
    C. Missouri control joint
    D. Tongue-and-groove joint

37. _____ *True or False?* Intersecting walls should be joined together in a masonry bond.

38. _____ Nonloadbearing walls can be tied to other walls using strips of metal lath or _____.
   A. building paper
   B. hardware cloth
   C. tie bars
   D. Z bars

39. _____ A(n) _____ plate is often anchored to the top of a concrete block wall to attach future framing.
   A. steel
   B. wood
   C. masonry
   D. anchor

40. _____ Concrete block cavity walls are constructed with two wythes separated by a continuous air space _____ wide.
   A. 1″ to 3 1/2″
   B. 1 1/2″ to 4″
   C. 2″ to 4 1/2″
   D. 2 1/2″ to 5″

41. _____ In a cavity wall, _____ is used to direct water to the weep holes.
   A. flashing
   B. cotton sash cord
   C. mortar
   D. continuous joint reinforcement

42. _____ *True or False?* The procedure for laying a 10″ concrete block cavity wall involves laying the first course of the inside wythe first, followed by the first course of the outside wythe.

43. _____ When laying a 10″ concrete block cavity wall, a recommended method of keeping the cavity free of mortar is to rest a(n) _____ on the joint reinforcement.
   A. metal tie
   B. board
   C. plastic cover
   D. unused block

44. _____ In the construction of a composite wall, _____ are often used as backup for bricks.
   A. concrete blocks
   B. wooden plates
   C. rigid insulation
   D. None of the above.

45. _____ A specially shaped _____ is used in composite walls greater than 10″ in thickness to bond the facing headers and backup units with sixth-course bonding.
   A. joint
   B. tie
   C. header block
   D. All of the above.

46. Piers are commonly used as support members for _____.

_____

47. _____ Which of the following should *not* be used to remove mortar droppings from a concrete block wall?
   A. Acid
   B. Trowel
   C. Putty knife
   D. Rubbing stone

48. _____ *True or False?* It is best practice to remove all mortar droppings from a concrete block wall while the mortar is still wet.

Name _____ Date _____ Class _____

CHAPTER 14 Stonemasonry

Carefully read Chapter 14 of the text and answer the following questions.

1. _____ *True or False?* Over the years, stone has declined in importance as a solid wall building material.

2. Mortar can be loaded on a trowel from the side of the board, the middle of the board, or the _____ of the pile.

_____

3. _____ The type of mortar and its _____ most often determine the speed of the stroke.
   A. brand
   B. color
   C. consistency
   D. intended use

4. _____ Mortar used for setting stone may be slightly _____ than mortar used for typical masonry units due to the weight of the pieces.
   A. darker
   B. lighter
   C. stiffer
   D. thinner

5. _____ Mortar on a board should be kept well _____ with water until it is used.
   A. tempered
   B. soaked
   C. protected
   D. stiffened

6. _____ *True or False?* Stone used for rubble or roughly squared stonework is usually fieldstone.

7. _____ Stones that are larger than average should be used near the _____ of a wall.
   A. bottom
   B. left side
   C. right side
   D. top

8. _____ A device, such as a sling or chain, that is attached to stones to lift them from the delivery trailer is called _____.
   A. a tagline
   B. rigging
   C. fall protection
   D. Both B and C.

9. _____ The person who attaches the rigging device to the stone to be lifted is called a(n) _____.
   A. lifter
   B. handler
   C. operator
   D. rigger

10. _____ The term for a safety rope or cable that is attached to the material being lifted so the person guiding the stone is a safe distance away from the load is called a _____.
    A. harness
    B. tagline
    C. forklift
    D. rigger

Copyright Goodheart-Willcox Co., Inc.
May not be reproduced or posted to a publicly accessible website.

11. _____ Which of the following statements is *false*?
    A. Safety slings should be kept in place until stones are within 6′ of their final location.
    B. Smooth-finish stones should be leaned face-to-face and back-to-back.
    C. Sand can be piled on projecting courses to protect them against mortar droppings.
    D. When working from scaffolding without handrails at or above 6′, workers must wear fall protection equipment.

12. _____ Which of the following is the correct ratio of mixture of materials for mortar used in stonemasonry?
    A. 1 part nonstaining cement, 1 part hydrated lime, 3 parts sand.
    B. 1 part nonstaining cement, 1 part hydrated lime, 6 parts sand.
    C. 1 part nonstaining cement, 3 parts hydrated lime, 6 parts sand.
    D. 1 part nonstaining cement, 6 parts hydrated lime, 6 parts sand.

13. _____ Low-alkali cement, called _____ cement, is recommended for setting limestone.
    A. stone
    B. nonstaining
    C. hydrated
    D. white

14. _____ *True or False?* Fieldstone and river rock *cannot* be used in their natural state.

15. _____ Splitting a stone with a _____ structure involves chipping on a marked line with the chisel end of a mason's hammer until a crack begins to develop.
    A. flat
    B. layered
    C. stratified
    D. Both B and C.

16. _____ A specialized hammer used to strike a chisel and designed for striking hardened steel is called a _____.
    A. brick hammer
    B. mason's hammer
    C. maul
    D. None of the above.

17. _____ Uncut stone or stone that has not been cut into a rectangular shape is called _____.
    A. ashlar
    B. limestone
    C. rubble
    D. trimmings

18. _____ When stone is precut with enough uniformity to allow some regularity in assembly, the wall is generally called _____.
    A. ashlar
    B. rubble
    C. trimmings
    D. veneer

Name _____

19. Review the following images and identify which image shows coursed ashlar and which shows random ashlar.

 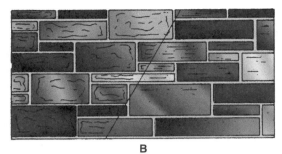

A   B

Indiana Limestone Institute

A. _____

B. _____

20. _____ Stones that are cut on all sides to specific dimensions are called _____.
    A. coursed ashlar
    B. points
    C. rubble
    D. trimmings

21. _____ *True or False?* When laying a rough stonemasonry veneer wall using rubble, enough nonstaining mortar should be mixed to last for one hour of work.

22. _____ When preparing to lay a rough stonemasonry veneer wall using rubble, select _____ stones for the bed course.
    A. small rubble
    B. large rubble
    C. large ashlar
    D. large panel

23. _____ Placing plastic mortar into raked-out joints is called _____.
    A. raking
    B. parging
    C. trimming
    D. pointing

24. _____ The ties used in successive courses of stone in a rough stonemasonry veneer wall should be _____.
    A. metal
    B. corrugated
    C. noncorrosive
    D. All of the above.

25. _____ *True or False?* Stone walls *cannot* be waterproofed.

26. _____ Walls that use mortar as a bond between stones require a footing below the _____ line because frost heaving can cause shifting in the wall structure.
    A. header
    B. base
    C. frost
    D. None of the above.

27. _____ Stone to be dry laid for a solid stone wall should have a base made of _____.
    A. gravel
    B. soil
    C. mortar
    D. rubble

28. _____ A solid stone wall depends on _____ stones for strength, stability, and tying the wall together.
   A. reinforced
   B. bond
   C. trimmed
   D. heavy

29. _____ The usual wall thickness of a solid stone wall is a minimum of _____.
   A. 14″
   B. 16″
   C. 18″
   D. 20″

30. _____ *True or False?* Limestone panels may be as thin as 1″.

31. _____ *True or False?* Quirk joints are less expensive on smooth finishes, while butt joints are less expensive on textured finishes.

32. Review the following image and identify which joint is a butt joint and which is a quirk joint.

   Indiana Limestone Institute

   A. _____
   B. _____

33. _____ Physical contact between the back face of stone walls and columns, slabs, and the ends of beams should be avoided because _____ can push the stone out.
   A. efflorescence
   B. expansion
   C. spalling
   D. deflection

34. _____ When installation below grade is unavoidable, either the trench method or the _____ method can be used to protect limestone panels from contact with soil.
   A. coating
   B. gravel
   C. dampproofing
   D. Both B and C.

35. _____ There should be a _____ minimum clearance between stone and all structural members.
   A. 1″
   B. 2″
   C. 3″
   D. 4″

36. _____ *True or False?* A strong pointing mortar usually performs better than a lean one.

Name _____

37. _____ In order to sustain the weight, an adequate number of _____ pads should be placed in the horizontal joint under heavy stones.
    A. steel
    B. support
    C. concrete
    D. setting

38. _____ The term _____ generally refers to the straps, rods, dovetails, and other connections that are embedded between the stone and the structure.
    A. *tie*
    B. *fastener*
    C. *anchor*
    D. *sealant*

39. _____ *True or False?* All anchors in limestone should be made of noncorrosive metal.

40. _____ A small cone-shaped device used to dispense mortar in joints where mortar is difficult to place using a trowel is called a mortar _____.
    A. bag
    B. case
    C. funnel
    D. shaft

41. _____ Anchor holes may be filled with any of the following, *except* _____.
    A. joint sealant
    B. expanding-type grout
    C. shim stock of a noncorrosive material
    D. mortar

42. _____ _____ cut stone after setting, rather than full bed setting and finishing in one operation, reduces settling due to shrinkage of the mortar bed.
    A. Supporting
    B. Raking
    C. Pointing
    D. Cleaning

43. _____ *True or False?* The bond strength of mortar is considerably reduced when mortar is frozen prior to hardening.

44. _____ The chemical reaction between water and cement that sets and hardens the cement is called _____.
    A. cementation
    B. hydration
    C. saturation
    D. solidification

45. _____ Which of the following statements about sealant systems is *false*?
    A. One-part systems are called moisture-cure or air-cure systems.
    B. Two-part systems rely on a catalyst or chemical agent to cure.
    C. Some sealant systems require a primer.
    D. Joints using a sealant system work best when the sealant is required to adhere to the backer rod.

46. _____ *True or False?* Limestone products have a high coefficient of expansion.

47. _____ In exterior limestone walls, _____ joints should be provided to reduce the damaging effect of thermal expansion of the building frame.
    A. expansion
    B. extension
    C. false
    D. sealant

48. _____ The best location for an expansion joint is in a(n) _____ of a building.
    A. frame
    B. flashing
    C. offset
    D. cornice

49. _____ In a typical cornice detail, the bed joint below is left _____ back far enough to remove any compressive stress that would have a tendency to break off stone below.
    A. anchored
    B. closed
    C. open
    D. supported

50. _____ *True or False?* Acid, wire brushes, and sandblasting are all usually permitted on stonework.

Name _____ Date _____ Class _____

CHAPTER 15
Foundation Systems

Carefully read Chapter 15 of the text and answer the following questions.

1. _____ The substructure of a building that resists settling and supports the weight of the building is called the _____.
    A. footing
    B. foundation
    C. pier
    D. slab

2. _____ The base on which a foundation is built is called the _____.
    A. footing
    B. slab
    C. excavation
    D. reinforcement

3. _____ Footings are usually _____ than the foundation wall to provide the extra support needed to stop or reduce uneven settling.
    A. narrower
    B. heavier
    C. taller
    D. wider

4. There are two general types of foundations, but only _____ foundations generally concern the work of masons.
   _____

5. _____ Footings that carry light loads and are not reinforced with steel are called _____ footings.
    A. plain
    B. light
    C. slab
    D. raft

6. _____ Which of the following is *not* a type of plain footing?
    A. Isolated
    B. Stepped
    C. Combined
    D. Continuous

7. _____ Steel is embedded in _____ footings to strengthen them.
    A. combined
    B. reinforced
    C. plain
    D. stepped

8. _____ As a rule of thumb, the depth of a residential footing placed on average soil should be _____.
    A. twice the thickness of the wall
    B. twice the thickness of the wall or a minimum of 8″
    C. slightly less than the thickness of the wall
    D. equal to the thickness of the wall or a minimum of 6″

For questions 9–12, match the footings shown in the following images with their correct type.

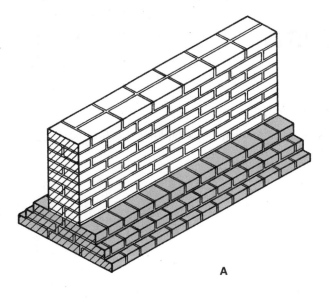
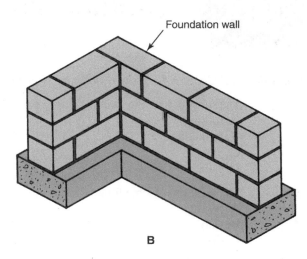
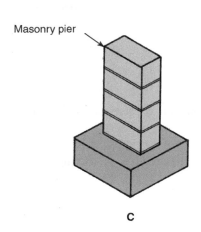
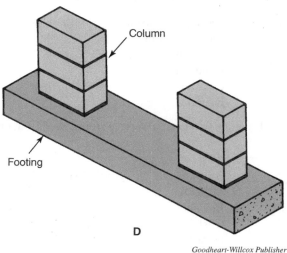

9. _____ Isolated footing

10. _____ Combined footing

11. _____ Continuous footing

12. _____ Stepped footing

13. _____ To aid in attaching a foundation wall to the footing, a recess or groove called a _____ is sometimes used to secure the two together.
    A. channel
    B. keyway
    C. notch
    D. cove

14. _____ The most common type of foundation wall is the _____.
    A. H foundation
    B. L foundation
    C. I foundation
    D. T foundation

Name _____

15. Identify the foundation types shown in the following image.

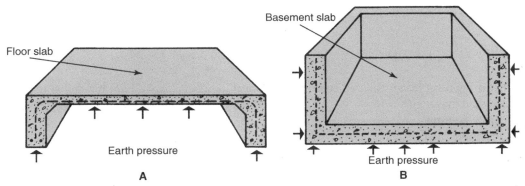

Goodheart-Willcox Publisher

A. _____

B. _____

16. _____ Foundation walls of hollow concrete block should be _____ with a course of solid masonry to help distribute the loads from floor beams and serve as a termite barrier.
    A. extended
    B. capped
    C. supported
    D. None of the above.

17. _____ When dampproofing, the outside of masonry basement walls should be parged with a _____-thick coating of Portland cement plaster or mortar.
    A. 1/4″
    B. 1/2″
    C. 3/4″
    D. 1″

18. _____ Creating a plastic cove is done by placing plaster on the top of the footing where the wall meets the _____.
    A. foundation
    B. stud
    C. footing
    D. soil

19. _____ *True or False?* Concrete walls can be dampproofed by spraying a sealant similar to tar on the wall and footing surfaces.

20. _____ Waterproofing involves the installation of perforated _____ around the footing and under the basement floor to remove accumulated water from around and under the foundation.
    A. gravel
    B. drain tiles
    C. keyways
    D. brick

21. _____ Drain tiles are laid in a bed of _____, and the area is later backfilled with _____.
    A. gravel; gravel
    B. gravel; soil
    C. soil; gravel
    D. mortar; gravel

22. _____ *True or False?* Columns and piers are built into the foundation wall.

23. The height of a column is at least _____ times its thickness.

24. _____ According to the MSJC Code and Specification, columns must be reinforced with a minimum of _____ reinforcing bars.
    A. two
    B. three
    C. four
    D. six

25. _____ What is the term for an isolated column of masonry or a bearing wall *not* bonded at the sides to associated masonry?
    A. Brick column
    B. Partition
    C. Pier
    D. Pilaster

26. _____ *True or False?* A pier can be used to support openings in a wall.

27. A pilaster may be bonded to create integral construction by interlocking _____% of the masonry units.
    _____

28. _____ _____ is the creation of temporary wall terminations of masonry units.
    A. Reinforcing
    B. Interlocking
    C. Anchoring
    D. Toothing

29. _____ When pilasters are built with concrete block, the hollow cells are sometimes filled with _____ to increase their strength.
    A. mortar
    B. gravel
    C. grout
    D. Both A and C.

30. _____ The building superstructure should be _____ the foundation to resist high winds.
    A. anchored to
    B. embedded to
    C. below
    D. above

31. _____ Sill plates of wood joint floor systems are generally anchored to masonry walls with 1/2″ bolts extending at least _____ into the filled cores of the masonry units.
    A. 1/2″
    B. 6″
    C. 15″
    D. 16″

32. _____ Anchor bolts should not be spaced more than _____ apart, with one bolt not more than 12″ from each end of the sill plate.
    A. 4′
    B. 8′
    C. 12′
    D. 15′

33. _____ Sill plates should be anchored to poured concrete walls with 1/2″ anchor bolts embedded _____, with the same maximum spacing as used for masonry walls.
    A. 6″
    B. 9″
    C. 12″
    D. 15″

Name _____ Date _____ Class _____

CHAPTER 16
Wall Systems

Carefully read Chapter 16 of the text and answer the following questions.

1. The heat flow through a building product based on the difference between external and internal temperatures is called the _____-value.

   _____

2. _____ Insulating materials are rated on their ability to resist heat loss, or _____.
   A. U-value
   B. R-value
   C. thickness
   D. thermal mass

3. _____ *True or False?* Masonry walls can be single-wythe or multiple-wythe.

4. _____ The characteristic of heat capacity and surface area capable of affecting building thermal loads by storing heat and releasing it at a later time is called _____ mass.
   A. heat
   B. building
   C. temperature
   D. thermal

5. _____ *True or False?* All solid masonry walls are loadbearing walls.

6. _____ The structural bond of solid masonry walls is provided by metal ties, _____, or joint reinforcement.
   A. masonry headers
   B. mortar
   C. anchors
   D. All of the above.

7. _____ The masonry units most commonly used for 6″ masonry wall construction are _____ bricks.
   A. ABI
   B. BIA
   C. RSC
   D. SCR

8. _____ A wall that is not wholly supported at each story is called a _____ wall.
   A. panel
   B. curtain
   C. cavity
   D. loadbearing

9. _____ A wall that is supported at each story and self-supporting between stories is called a _____ wall.
   A. panel
   B. blind
   C. curtain
   D. structural

10. _____ The horizontal joint reinforcement in a 4″ wall resists _____ stresses that result from _____ pressures as the wall spans horizontally between structural elements.
    A. compressive; lateral
    B. horizontal; lateral
    C. tensile; lateral
    D. tensile; vertical

11. Ladder- or _____-type joint reinforcement provides two longitudinal bars connected by cross wires.

_____

12. _____ *True or False?* Water penetration in a 4″ wall can be controlled by providing drainage space or applying a water barrier to the inside of the wall.

13. _____ Since 4″ walls are considered to span horizontally, any expansion joints must be located at _____ supports.
    A. inside
    B. horizontal
    C. outside
    D. vertical

14. _____ Hollow masonry walls are built using _____ or hollow masonry units.
    A. solid
    B. bonded
    C. plastic
    D. steel

15. _____ The exterior wythe of a cavity wall is usually a nominal _____ thick.
    A. 2″
    B. 3″
    C. 4″
    D. 5″

16. _____ The interior wythe of a cavity wall can be _____ thick.
    A. 2″, 4″, or 6″
    B. 4″, 6″, or 8″
    C. 6″, 8″, or 10″
    D. 8″, 10″, or 12″

17. _____ A mortar opening that creates a void where moisture inside the wall cavity can exit to the outside is called _____.
    A. a water hole
    B. a cold joint
    C. a weep hole
    D. flashing

18. _____ _____ is a thin, impervious material placed in mortar joints that keeps any moisture that may collect in the cavity away from the inner wall.
    A. Flashing
    B. Insulation
    C. Veneer
    D. None of the above.

19. _____ *True or False?* Heat losses and heat gains through masonry walls are maximized by the use of cavity wall construction.

20. _____ Sound resistance in a cavity wall is accomplished with heavy, massive walls or _____ construction.
    A. continuous
    B. discontinuous
    C. isolated
    D. Both A and C.

21. _____ Fire resistance ratings of cavity walls range from _____ hour(s) to _____ hours.
    A. 1, 3
    B. 2, 4
    C. 2, 5
    D. 3, 5

22. _____ *True or False?* Rigid boards and granular fills are suitable types of insulation for cavity walls.

23. _____ *True or False?* The minimum amount of space that should be left between the cavity face of the external wythe and the insulation board is 1″.

Name _____

24. _____ Adjacent wythes of cavity walls should be tied together using at least one metal tie for each _____ sq ft of wall area.
    A. 3
    B. 3 1/2
    C. 4
    D. 4 1/2

25. _____ Expansion joints are recommended through outer wythe walls of cavity walls where the walls are _____ feet or more in length.
    A. 10'
    B. 25'
    C. 50'
    D. 60'

26. _____ *True or False?* In anchored veneer walls, facing veneer is attached to the backing and acts structurally with the rest of the wall.

27. _____ In anchored brick veneer construction, a nominal _____-thick exterior brick wythe is anchored to a backing system with metal ties.
    A. 1" or 2"
    B. 2" or 3"
    C. 3" or 4"
    D. 4" or 5"

28. _____ *True or False?* A veneer wall has a facing of masonry units or other weather-resisting, noncombustible materials.

29. _____ Brick veneer wall assemblies are _____-type walls that are resistant to rain penetration.
    A. drainage
    B. waterproof
    C. flashing
    D. support

30. _____ *True or False?* Brick veneer with wood or metal frame backing is usually built with a minimum of 2" of air space.

31. _____ Brick veneer on a frame backing transfers the weight of the veneer to the _____.
    A. footing
    B. foundation
    C. adjacent wall
    D. backing

32. _____ In a masonry wall with wood backing, there should be one tie for every 2 2/3 sq ft of wall area, with a maximum spacing of _____ O.C. in either direction.
    A. 12"
    B. 18"
    C. 24"
    D. 30"

33. _____ In a masonry wall with wood backing, corrugated ties must penetrate to at least _____ the veneer thickness and at least 5/8" mortar cover.
    A. double
    B. triple
    C. half
    D. a third of

34. _____ Flashing and weep holes must be positioned above grade but as close to the _____ of the wall as possible.
    A. bottom
    B. left
    C. right
    D. top

35. _____ Weep holes in which the mortar has been completely removed should be spaced not more than _____ O.C.
   A. 16″
   B. 18″
   C. 20″
   D. 24″

36. _____ Weep holes with wicking material should be spaced at a maximum of _____ O.C.
   A. 16″
   B. 18″
   C. 20″
   D. 24″

37. _____ Unless the masonry is self-supporting, brick veneer that is backed by wood or metal frame must be supported by _____ over openings.
   A. grout
   B. mortar
   C. lintels
   D. corrugated ties

38. _____ Expansion joints may be required in brick veneer when there are long walls, walls with returns, or large _____.
   A. loads
   B. openings
   C. bricks
   D. frames

39. _____ Walls with two wythes bonded together with masonry or wire ties are called _____ walls.
   A. cavity
   B. composite
   C. hollow
   D. solid

40. _____ The narrow space between the facing units and the backup units in a composite wall is called a(n) _____ joint.
   A. collar
   B. expansion
   C. header
   D. weep hole

41. _____ *True or False?* The first course of facing in an 8″ composite wall may be either headers or stretchers.

42. _____ *True or False?* The vapor pressure differential across a wall section can be decreased through ventilation and dehumidification.

43. _____ A mechanical device that reduces heat loss and exchanges moisture-laden indoor air with outside air is called a _____.
   A. heater/dehumidifier
   B. heat exchanger
   C. vapor barrier
   D. vapor exchanger

44. _____ Walls built with steel reinforcement embedded in the masonry units are called _____ masonry walls.
   A. composite
   B. hollow
   C. reinforced
   D. grouted

45. _____ *True or False?* Maximum spacing of principal reinforcement should not exceed 24″.

46. _____ For walls over 3′ in height, the _____-type retaining wall provides the best solution.
   A. cantilever
   B. gravity
   C. grout
   D. garden

Name _____

47. _____ In grouted masonry walls, grout is added to the cores in loadbearing masonry walls to provide added _____.
   A. stability
   B. weather resistance
   C. corrosion resistance
   D. strength

48. _____ For straight garden walls, it is recommended that for 10 psf wind pressure, the height above grade not exceed _____ of the wall thickness squared.
   A. 1/2
   B. 3/4
   C. 3/8
   D. 5/8

49. _____ _____ walls are ideal for uneven terrain.
   A. Gravity
   B. Straight garden
   C. Pier and panel
   D. Serpentine

50. _____ *True or False?* The radius of curvature for a 4″ serpentine wall should be no more than half the height of the wall above the grade.

51. _____ Thin brick veneer is approximately _____ thick.
   A. 1/4″ to 3/4″
   B. 1/2″ to 1″
   C. 3/4″ to 1 1/4″
   D. 1″ to 1 1/2″

52. _____ In the thin bed set installation procedure, an epoxy or organic adhesive is typically used on _____ surfaces only.
   A. interior
   B. exterior
   C. flat
   D. curved

53. _____ _____ are usually constructed indoors in a factory where they cure under ideal conditions.
   A. Hardscapes
   B. Precast panels
   C. Mortar and grout mixes
   D. Retaining walls

54. _____ Lip locks on concrete masonry units are used to provide proper alignment and prevent _____.
   A. sagging
   B. forward movement exerted by earth pressure
   C. corrosion of units
   D. tilting of units during layout

55. Segmental retaining walls are constructed of high-strength concrete blocks or units made specifically for _____ stacking.

_____

56. _____ Retaining walls and other masonry landscaping materials are sometimes called _____.
   A. composites
   B. hollow units
   C. hardscape
   D. None of the above.

57. _____ Who should be contacted to determine the local frost line depth?
   A. Local building code office.
   B. Federal building code office.
   C. Job supervisor.
   D. Local contractor.

58. _____ *True or False?* When installing a retaining wall, the drain tile should be installed behind either the first or second course.

59. _____ Spaces left in a wall for the purpose of containing plumbing, heating ducts, electrical wiring, or other equipment can be in the form of _____.
   A. chases
   B. weep holes
   C. recesses
   D. Both A and C.

60. _____ What is the term for a structural member placed over an opening in a wall to support the loads above that opening?
   A. Brace
   B. Joist
   C. Lintel
   D. Strut

61. _____ *True or False?* Steel lintels should be supported on either side of the opening for a distance of at least 6″.

62. _____ Slight movement often occurs at the location of the lintels. To account for this movement, _____ joints are often located at the ends of lintels.
   A. control
   B. cold
   C. expansion
   D. construction

63. _____ Masonry structures that span an opening by transferring vertical loads laterally to adjacent masonry units, and thus to abutments, are called _____.
   A. arches
   B. bridges
   C. lintels
   D. skewbacks

64. _____ The surface on which an arch joins the supporting abutment is called the _____.
   A. joint
   B. skewback
   C. soffit
   D. span

Name _____

65. Identify the types of arches shown in the following images.

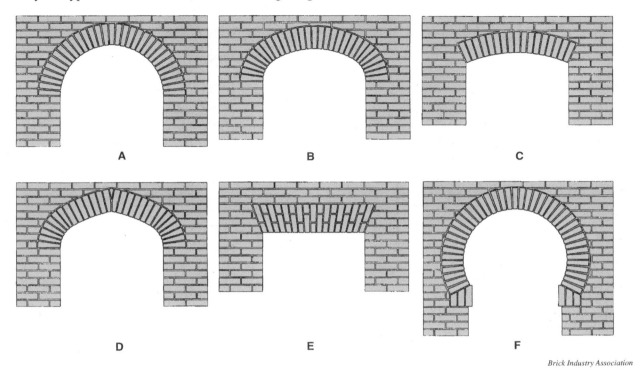

Brick Industry Association

A. _____

B. _____

C. _____

D. _____

E. _____

F. _____

66. Review the following image and identify the terms for the components of the arch shown.

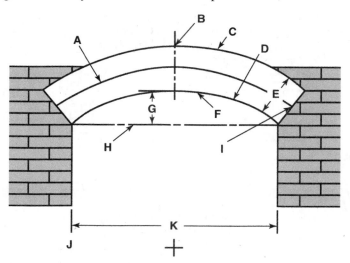

A. _____   G. _____

B. _____   H. _____

C. _____   I. _____

D. _____   J. _____

E. _____   K. _____

F. _____

67. _____ The masonry unit located at the center of the crown of the arch is called the _____.
   A. cap
   B. headstone
   C. keystone
   D. peak

68. _____ Single masonry units typically laid in radial orientation for arches are called _____.
   A. keystones
   B. voussoirs
   C. caps
   D. extrados

69. _____ The dimension of an arch at the skewback that is perpendicular to the arch axis (except in jack arches) is called the _____.
   A. rise
   B. skewback
   C. depth
   D. spring line

70. _____ Arches are constructed with the aid of temporary shoring, or _____, which supports the masonry while the arch is being constructed.
   A. bracing
   B. centering
   C. holding
   D. joisting

80 Modern Masonry Lab Workbook

Name _____

71. _____ Which of the following statements is *false*?
    A. A jack arch should be supported by steel if the opening is over 2′ wide.
    B. Each joint in a jack arch should be the same width as the entire length of the joint for best appearance.
    C. The most important features to be considered when constructing a segmental arch are the depth and camber.
    D. The rise of a segmental arch should be 1/6, 1/8, 1/10, or 1/12 of the span.

72. _____ The _____ are the vertical sides of a door or window.
    A. heads
    B. jambs
    C. sills
    D. arches

73. _____ The purpose of a window _____ is to channel water away from the building.
    A. lintel
    B. head
    C. sill
    D. ledge

74. _____ Joints that separate brick masonry into segments to prevent cracking due to changes in temperature, moisture expansion, elastic deformation due to loads, and creep are called _____ joints.
    A. control
    B. construction
    C. cold
    D. expansion

75. _____ A joint that creates a plane of weakness in concrete masonry to control the location of cracks is called a(n) _____ joint.
    A. expansion
    B. control
    C. building expansion
    D. cold

76. _____ Joints that divide a building into separate sections so stresses developed in one section do not affect the integrity of the rest of the structure are called _____ joints.
    A. building expansion
    B. control
    C. construction
    D. expansion

77. _____ A joint that is used in concrete construction where masonry work is interrupted is called a _____ joint.
    A. building expansion
    B. control
    C. construction
    D. partition

78. Because different masonry products expand and contract differently under stress and temperature, _____ breaks are necessary to prevent cracks.

    _____

79. _____ *True or False?* Quoins are large squared stones used at corners and around openings of buildings for ornamental purposes.

80. _____ A covering within the height of the wall, normally where there is a change in wall thickness, is called a _____.
    A. cap
    B. coping
    C. guard
    D. shield

81. _____ What is the correct term for the covering placed at the top of a wall?
   A. Cap
   B. Coping
   C. Crest
   D. Crown

82. _____ A shelf or ledge formed by projecting successive courses of masonry out from the face of a wall is called a _____.
   A. cap
   B. corbel
   C. racking
   D. coping

83. _____ What is the correct term for masonry in which successive courses are stepped back from the face of the wall?
   A. Framing
   B. Segmenting
   C. Racking
   D. Staging

Name _____ Date _____ Class _____

Chapter 17: Paving and Masonry Construction Details

Carefully read Chapter 17 of the text and answer the following questions.

1. _____ Bricks that are fired in a kiln at high temperatures to increase their mechanical strength, moisture resistance, and weather resistance are called _____ bricks.
    A. hard-burned
    B. solid-fired
    C. stone-fired
    D. All of the above.

2. Brick that has an average saturation coefficient of _____ or less is recommended for severe weathering conditions.

   _____

3. _____ *True or False?* Bricks laid flat with the largest plane surfaces horizontal are generally cored bricks.

4. _____ *True or False?* Interior masonry floors are usually laid over concrete slabs.

5. _____ The procedure for laying a masonry floor on a slab begins with choosing a _____.
    A. material
    B. pattern
    C. joint
    D. cushion

6. _____ When laying paving brick, edge _____ should be installed to keep the units from shifting once the paving bricks are installed.
    A. ties
    B. bars
    C. restraints
    D. patterns

7. _____ The slope of walks and patios is generally _____ per foot.
    A. 1/8″ to 1/4″
    B. 1/6″ to 1/2″
    C. 1/4″ to 3/4″
    D. 1/2″ to 1″

8. _____ Below-surface drainage may be required for large paved areas or locations with a high _____.
    A. soil table
    B. elevation
    C. frost line
    D. water table

9. _____ Paver bricks are _____.
    A. designed especially for paving
    B. approximately 1″ thinner than regular bricks
    C. 1 1/4″ × 3 5/8″ × 7 5/8″ in size
    D. All of the above.

10. _____ When steps are built with masonry units, a concrete foundation is poured for the structure, and the treads and _____ are capped with the material selected.
    A. joints
    B. risers
    C. edges
    D. Both A and C.

11. _____ Treads for all outside steps are usually _____.
    A. 10″
    B. 12″
    C. 14″
    D. 16″

12. _____ *True or False?* Treads should be sloped about 1/2″ to the front edge to aid in water drainage.

13. _____ A room with 300 sq ft of floor area can be adequately served by a fireplace with an opening _____ wide.
    A. 20″ to 26″
    B. 25″ to 31″
    C. 30″ to 36″
    D. 35″ to 41″

14. _____ The space or area where combustion occurs to create heat in the fireplace is called the combustion _____.
    A. cavity
    B. chamber
    C. compartment
    D. shelf

15. _____ The draft in a fireplace is affected by the size of the opening, the combustion chamber, and the _____.
    A. flue
    B. chimney
    C. room where the fireplace is built
    D. firebox

16. _____ The International Residential Code requires a depth of _____ for the firebox of a masonry fireplace.
    A. 8″
    B. 12″
    C. 20″
    D. 36″

Name _____

Review the image shown and answer questions 17–24, identifying the different components of a firebox assembly using the appropriate letter.

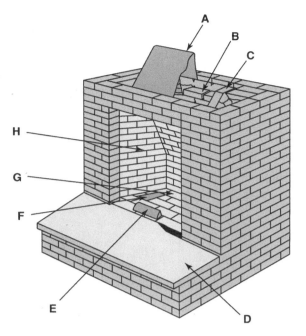

Goodheart-Willcox Publisher

17. _____ Air inlet damper

18. _____ Combustion chamber

19. _____ Throat

20. _____ Raised hearth

21. _____ Inner hearth

22. _____ High formed damper

23. _____ Smoke shelf

24. _____ Ash dump

25. _____ The back and end walls of an average-size fireplace are generally _____ thick.
    A. 6″
    B. 8″
    C. 10″
    D. 12″

26. _____ *True or False?* A narrow passage in a fireplace that connects the firebox with the smoke chamber is called the *channel*.

27. _____ A _____ closes when the chimney is not in use to reduce air infiltration down the chimney.
    A. firebox
    B. throat
    C. channel
    D. damper

28. _____ The smoke _____ is located directly above the firebox and compresses gases created from burning wood or other flammable materials.
    A. chamber
    B. plate
    C. compressor
    D. shelf

Chapter 17 Paving and Masonry Construction Details 85

29. _____ *True or False?* A chimney flue lining is usually a fired clay pipe.

30. _____ Which of the following statements about hearths is *false*?
    A. Hearths are usually built using a reinforced brick masonry cantilevered slab.
    B. A reinforced concrete slab may be used to construct a hearth.
    C. According to the IRC, hearths must be at least 6″ thick.
    D. The opening for the ash chute is formed in the hearth.

31. _____ The foundation wall for a masonry chimney should be a minimum of _____ thick.
    A. 4″
    B. 6″
    C. 8″
    D. 12″

32. _____ Changes to the size and shape of a chimney must not be made within _____ above or below the roof joists or rafters.
    A. 2″
    B. 4″
    C. 6″
    D. 8″

33. Chimney walls must be at least _____″ thick for residential structures.

_____

34. _____ Chimney walls must be lined with fire clay flue liners not less than _____ thick.
    A. 1/4″
    B. 3/8″
    C. 1/2″
    D. 5/8″

35. _____ *True or False?* When a chimney has two flues not separated by masonry, the joints of adjacent flue linings should be staggered at least 7″.

36. _____ Flashing and counterflashing should be made of a _____-resistant material, such as copper.
    A. weather
    B. wind
    C. rust
    D. None of the above.

37. _____ Prefabricated steel heat circulating fireplaces are efficient because the _____ where air is heated and returned through vents.
    A. sides and back are double-walled
    B. back is single-walled and sides are double-walled
    C. sides are single-walled and back is double-walled
    D. back and sides are single-walled

38. _____ The type of fireplace that joins an almost airtight set of doors on the 1/4″ plate steel firebox is the _____.
    A. brick open steel firebox
    B. masonry enclosed steel firebox
    C. prefabricated steel heat circulating fireplace
    D. None of the above.

39. _____ The inside wall of masonry enclosed steel firebox fireplaces is usually made of _____ masonry block.
    A. concrete
    B. solid
    C. hollow
    D. Haydite

40. _____ *True or False?* When a masonry enclosed steel firebox fireplace is in use, the doors should be left open to improve energy efficiency.

Name _____ Date _____ Class _____

CHAPTER 18
Concrete Materials and Applications

Carefully read Chapter 18 of the text and answer the following questions.

1. _____ Ingredients added to the basic concrete mix in order to produce certain properties in the finished product are called _____.
   A. admixtures
   B. fine aggregates
   C. coarse aggregates
   D. All of the above.

2. _____ Portland cement is _____, which means it sets and hardens by reacting with water.
   A. pneumatic
   B. hydraulic
   C. tempered
   D. air-entraining

3. _____ Which of the following is a general-purpose cement that is suitable for all applications?
   A. Type I Portland cement
   B. Type II Portland cement
   C. Type III Portland cement
   D. Type IV Portland cement

4. _____ Which of the following produces less heat due to hydration and can resist sulfate attack?
   A. Type II Portland cement
   B. Type III Portland cement
   C. Type IV Portland cement
   D. Type V Portland cement

5. _____ Cement that is used when forms must be removed as quickly as possible or when the structure is to be used quickly is _____ Portland cement.
   A. modified
   B. high early strength
   C. low-heat
   D. sulfate-resistant

6. _____ Cement that is designed for use in massive structures, such as dams when heat generated by hydration must be kept low, is called _____ Portland cement.
   A. air-entraining
   B. high early strength
   C. low-heat
   D. sulfate-resistant

7. _____ *True or False?* Type V sulfate-resistant Portland cement is used only in construction exposed to severe sulfate action.

8. _____ Cement that improves resistance to freeze-thaw action and scaling is called _____ Portland cement.
   A. air-entraining
   B. white
   C. low-heat
   D. sulfate-resistant

9. _____ Portland-pozzolan cements are used principally for _____.
   A. architectural purposes
   B. construction exposed to severe sulfate action
   C. large hydraulic structures
   D. structures that need to be used quickly

10. _____ Aggregates usually make up _____ of the volume of concrete.
    A. 40% to 60%
    B. 50% to 70%
    C. 60% to 80%
    D. 70% to 90%

11. Aggregates that vary uniformly in size from very fine up to 1/4″ in diameter are categorized as _____ aggregates.
_____

12. _____ As a rule of thumb, the maximum size of coarse aggregate is usually _____ in diameter.
    A. 1/4″
    B. 1/2″
    C. 1 1/4″
    D. 1 1/2″

13. _____ The primary purpose of adding water to cement is to cause _____ of the cement.
    A. hardening
    B. hydration
    C. blending
    D. None of the above.

14. _____ Concrete mixing water should be free of oil, _____, and acid.
    A. alkali
    B. sulfate
    C. sediment
    D. Both A and B.

15. _____ _____ admixtures improve the durability of concrete exposed to moisture during cycles of freezing and thawing.
    A. Water-reducing
    B. Air-entraining
    C. Retarding
    D. Accelerating

16. _____ The standard amount of air content in concrete commonly requested by the contractor is _____.
    A. 1% to 3%
    B. 3% to 5%
    C. 5% to 7%
    D. 7% to 9%

17. _____ Which of the following is used in air-entrainment testing?
    A. Electronic scale
    B. Tamping rod
    C. Rubber mallet
    D. All of the above.

18. _____ Materials that reduce the quantity of mixing water required to produce concrete of a given consistency are called _____ admixtures.
    A. air-entraining
    B. water-reducing
    C. retarding
    D. accelerating

Name _____

19. _____ _____ is the relative consistency or stiffness of plastic concrete.
    A. Shrinkage
    B. Fluidity
    C. Slump
    D. Thickness

20. _____ A(n) _____ admixture is used when it is desirable to offset rapid setting time of concrete due to hot weather.
    A. accelerating
    B. pozzolan
    C. retarding
    D. water-reducing

21. _____ *True or False?* A pozzolan is used to speed up the setting and strength development of concrete.

22. _____ The most commonly used accelerating admixture is _____.
    A. calcium chloride
    B. calcium hydroxide
    C. entrained air
    D. pozzolan

23. _____ A siliceous or siliceous and aluminous material that possesses little or no cementitious value is a(n) _____.
    A. sulfate
    B. alkali
    C. pozzolan
    D. retarder

24. _____ *True or False?* In properly proportioned concrete, each particle of aggregate is completely surrounded by the cement-water paste.

25. _____ The water-cement ratio is stated in _____ of water per bag (94 lb) of cement.
    A. liters
    B. gallons
    C. pounds
    D. quarts

26. _____ Only about _____ gallons of water per bag of cement is necessary to complete the chemical reactions between cement and water.
    A. 3 1/2
    B. 4
    C. 9
    D. 10

27. _____ After concrete is in place, the strength continues to increase as long as satisfactory _____ and temperatures are present.
    A. air movement
    B. dryness
    C. moisture
    D. Both A and B.

28. _____ A concrete compression testing machine operates as a press and registers the point at which concrete _____.
    A. sets
    B. hardens
    C. fractures
    D. bends

29. _____ *True or False?* Rule-of-thumb approximation of volumes, such as 1:2:3 of cement to sand to gravel, are recommended for quality concrete mixtures.

30. _____ Very fluid mixes are called _____ concrete.
    A. dry-slump
    B. high-slump
    C. low-slump
    D. wet-slump

31. Generally, a(n) _____-slump concrete produces a better concrete product.

_____

32. _____ A slump test can be performed in the field with a metal slump _____.
    A. cone
    B. rod
    C. rule
    D. All of the above.

33. _____ *True or False?* A slump test is a rough measure of the consistency of concrete but should *not* be considered a measure of workability.

34. _____ *True or False?* Volume is an accurate method of measuring cement.

35. _____ Measurement of aggregates by _____ is the recommended practice.
    A. consistency
    B. number of bags
    C. volume
    D. weight

36. _____ Generally, mixing time should be at least one minute for mixtures up to 1 cu yd with an increase of _____ seconds for each 1/2 cu yd or fraction thereof.
    A. 5
    B. 15
    C. 20
    D. 30

37. _____ ASTM C94 requires concrete to be delivered and discharged within _____ hour(s) of water being added to the mix.
    A. 1
    B. 1.5
    C. 2
    D. 2.5

38. _____ *True or False?* To make concrete more workable if it has dried out and stiffened after it has been left standing, it is acceptable to add water.

39. _____ A composite material in which the concrete resists compression forces and steel bars or wires resist tension forces is called _____ concrete.
    A. tensile
    B. durable
    C. reinforced
    D. resistant

40. _____ The best reinforcing material for concrete is _____.
    A. aluminum
    B. brass
    C. titanium
    D. steel

41. _____ Reinforcing bars can either be smooth or _____.
    A. corrugated
    B. galvanized
    C. deformed
    D. rough

Name _____

42. Cross-sectional area is the basic element used in specifying wire size; smooth wire sizes are specified by the letter ____, and deformed wires are specified by the letter *D*.

_____

43. _____ Steel reinforcement should be placed so it is protected by adequate coverage of _____.
    A. welded wire
    B. concrete
    C. cement
    D. mortar

44. _____ *True or False?* Reinforcing steel must be lapped at a splice.

45. The overlap distance for deformed bars should be at least ____ bar diameters.

_____

46. _____ *True or False?* A cubic foot of concrete is considered lightweight if it has a density or weight between 105 and 125 pounds per cubic foot.

47. _____ Using lightweight aggregates reduces the weight of concrete about _____ compared with the weight of concrete made with normal-weight aggregates.
    A. 20% to 45%
    B. 15% to 40%
    C. 10% to 35%
    D. 25% to 50%

Notes

CHAPTER 19 / Form Construction

Carefully read Chapter 19 of the text and answer the following questions.

1. _____ The most popular form material is _____.
   A. steel
   B. wood
   C. concrete
   D. plastic

2. _____ Which of the following statements about Plyform is *false*?
   A. It is specially designed for concrete forming.
   B. Panel size is 4′ × 8′.
   C. Class II Plyform is stronger and stiffer than Class I Plyform.
   D. Available thicknesses are 19/32″, 23/32″, and 3/4″.

3. The rate of bending plywood can be increased for a given thickness sheet by sawing _____ across the inner face at right angles to the curve.

   _____

4. _____ Hardboard that is tempered, specially treated, and _____ thick can be used as a form facing material.
   A. 1/4″
   B. 1/2″
   C. 3/4″
   D. 1″

5. _____ *True or False?* Steel and aluminum frames and form facings provide greater strength and support heavier loads than wood frames.

6. _____ Which form material is increasing in use and is frequently used in pan forms for ribbed concrete floors?
   A. Aluminum
   B. Fiberglass
   C. Plywood
   D. Steel

7. _____ *True or False?* Insulating board and rigid foam can be used as form liners.

8. _____ The primary consideration in form design is usually its _____.
   A. appearance
   B. cost
   C. height
   D. strength

9. View the following image and identify the different parts of the concrete form.

A. _____
B. _____
C. _____
D. _____
E. _____
F. _____
G. _____
H. _____

10. _____ The form element that gives the concrete surface its shape and texture is called the _____.
    A. brace
    B. sheathing
    C. stud
    D. wale

11. _____ The term for the form element that supports the sheathing and prevents bowing is _____.
    A. brace
    B. sheathing
    C. stud
    D. wale

12. _____ The base of the wall form that is temporarily attached to the footing is called the _____.
    A. sill
    B. brace
    C. sheathing
    D. tie

13. _____ The part of the concrete form that is used to align the forms, secure the ties, and support the studs are the _____.
    A. braces
    B. stakes
    C. ties
    D. wales

14. _____ A type of brace that is used to align the form and is only required on one side of the form is the _____.
    A. tie
    B. spreader
    C. turnbuckle
    D. Both A and B.

15. _____ Spreaders made of _____ are used to keep form walls separated until concrete is in place and then are removed as the concrete is poured to their level.
    A. fiberglass
    B. galvanized iron
    C. steel
    D. wood

94 Modern Masonry Lab Workbook

Name _____

16. _____ *True or False?* Metal or fiberglass ties remain in the concrete and become a permanent part of the structure.

17. _____ The depth that a tie is broken off in the concrete is called the _____.
    A. spread
    B. break back
    C. break depth
    D. thread depth

18. _____ The most common type of continuous single-member tie is the _____.
    A. snap tie
    B. loop and tie
    C. internal disconnecting tie
    D. None of the above.

19. _____ What type of tie is designed for heavier construction work?
    A. Snap
    B. Loop and tie
    C. Internal disconnecting
    D. Concrete form

20. _____ The most common type of prefabricated form is made of _____.
    A. a metal frame with a plywood facing
    B. a metal frame with an aluminum facing
    C. a wood frame with a plywood facing
    D. a wood frame with an aluminum facing

21. _____ Which of the following is the most common modular size of prefabricated forms?
    A. 1′ × 6′
    B. 1′ × 8′
    C. 2′ × 4′
    D. 2′ × 8′

22. _____ _____ forms are used for casting very large structures of great height.
    A. Cast
    B. Glide
    C. Gang
    D. Slip

23. _____ *True or False?* Slip forms are raised very quickly by jacks.

24. _____ Soil can serve as a form for footings if it is firm and not too _____.
    A. wet
    B. dry
    C. porous
    D. lumpy

25. _____ Concrete footing forms are usually constructed from _____ construction lumber the same width as the footing thickness.
    A. 1″
    B. 2″
    C. 3″
    D. 4″

26. _____ Footings for piers and columns that are formed with bottomless boxes made of construction lumber are called _____ footings.
    A. box
    B. isolated
    C. pack
    D. tapered

27. _____ This image is an example of a(n) _____ footing.
   A. isolated
   B. tapered
   C. stepped
   D. support

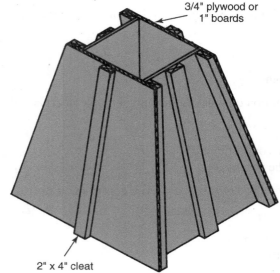

28. _____ What is the first step in building in-place forms?
   A. Cut the studs to the desired length.
   B. Attach the wales.
   C. Attach the sole plate to the footing.
   D. Toe-nail the wall studs.

29. _____ *True or False?* When building a wall form, you should increase stud spacing as the height of the wall increases.

30. _____ Any wall openings, such as for windows, can be formed using _____.
   A. braces
   B. wales
   C. bucks
   D. sheathing

31. _____ A typical buck design consists of an outside frame reinforced by 2 × 4s placed on edge with horizontal _____.
   A. wales
   B. cross braces
   C. box frames
   D. sheet metal sleeves

32. _____ *True or False?* On large projects, wood forms are generally used to form slabs.

33. _____ *True or False?* When a slab form is attached to stakes, the spacing recommended is about 2′ for 1″-thick formwork.

34. _____ Curves with a short radius can be formed with _____ plywood with the face grain vertical.
   A. 1/8″ or 1/4″
   B. 1/4″ or 1/2″
   C. 1/2″ or 3/4″
   D. 3/4″ or 1″

35. _____ Which of the following statements is *true*?
   A. A builder's transit can move in the vertical plane.
   B. Readings with a builder's level are taken in a horizontal plane.
   C. The proper grade can be maintained by using a mason's line or builder's level.
   D. All of the above.

Name _____

36. _____ Which of the following is an example of a permanent form?
    A. Divider strips in a patio
    B. Slab forms
    C. Screen tiles
    D. All of the above.

37. _____ A type of permanent form that casts a void in concrete slabs, beams, or other applications is a _____.
    A. cylindrical column
    B. hollow tube
    C. divider strip
    D. None of the above.

38. _____ Construction joints are formed by placing a _____ in the form.
    A. bulkhead
    B. divider
    C. spacer
    D. None of the above.

39. _____ *True or False?* One type of concrete steps can be built directly on the slope of the ground.

40. _____ Self-supporting steps must be reinforced with steel and rest on a(n) _____.
    A. foundation
    B. isolated footing
    C. sole plate
    D. slab

41. _____ Square and rectangular column forms are easily constructed from _____.
    A. aluminum
    B. fiberglass
    C. paper
    D. wood

42. _____ *True or False?* Round column forms that are made of fiberboard and paper are reusable.

43. _____ Which of the following is commonly used to maintain proper spacing between forms and rebars in form construction?
    A. Spacer blocks
    B. Snap-on devices
    C. Wood strips
    D. All of the above.

44. _____ A release agent that develops a film between the forming surface and the concrete to prevent adhesion is called a _____ release agent.
    A. reactive
    B. blocking
    C. barrier
    D. partition

45. _____ In order to avoid damaging the forms during stripping, _____ should be used.
    A. metal pry bars
    B. a hammer
    C. a wood wedge
    D. None of the above.

46. _____ Stay-in-place insulated forms are called _____.
    A. insulated concrete forms
    B. installed concrete forms
    C. insulated cavity forms
    D. None of the above.

47. _____ *True or False?* ICF wall construction requires rebars.

48. _____ In the top last course of an ICF wall, _____ are inserted when the cavity is filled with concrete.
   A. metal ties
   B. anchor bolts
   C. wood wedges
   D. form ties

49. _____ *True or False?* One of the disadvantages of the use of ICFs is that it increases the time needed to construct concrete walls.

50. _____ Which of the following statements about ICF walls is *false*?
   A. R-value is considerably higher than that of a concrete poured wall.
   B. The construction season can be extended in cold climates.
   C. Installation procedures do not vary among manufacturers.
   D. As the courses are stacked on each other, the walls need to be braced.

51. _____ An advantage of ICF construction when placed above grade is the _____.
   A. heavier weight of the forms
   B. lower cost of construction
   C. lower R-value of the wall
   D. sound resistance of the structure

52. _____ When constructing ICF walls, the installer creates cutouts called _____ by leaving out blocks to create voids in the wall.
   A. blockouts
   B. blowouts
   C. Fox Blocks
   D. Both A and B.

53. _____ *True or False?* The interior surface of an ICF must be covered with a thermal barrier to protect the insulation from fire.

54. _____ On the exterior of an ICF structure, exposed foam that comes in contact with backfilling materials must be _____.
   A. dampproofed
   B. dried
   C. removed
   D. sealed

55. _____ Concrete must not be placed too rapidly into an ICF structure or _____ will occur.
   A. blockout
   B. blowout
   C. break back
   D. bulking

56. _____ _____ are reversible, preassembled foam blocks designed to achieve a close tolerance when the structure is fabricated.
   A. Fox Blocks
   B. Lite-Forms
   C. Greenblocks
   D. SmartBlocks

57. _____ The _____ system is a foam concrete forming system that uses metric foam foundation units to form walls and foundations.
   A. Lite-Form
   B. Greenblock
   C. SmartBlock
   D. Fox Blocks

58. _____ *True or False?* SmartBlock polystyrene blocks are 10″ × 10″ × 40″ and weigh 2 lb each.

Name _____ Date _____ Class _____

CHAPTER 20
Concrete Flatwork and Formed Shapes

Carefully read Chapter 20 of the text and answer the following questions.

1. _____ Concrete can be cast in practically any shape as long as a _____ can be built to contain it while it is in a plastic state.
   A. barrier
   B. wall
   C. form
   D. support

2. _____ *True or False?* The consistency of the concrete used on a job is based on what the delivery equipment can handle.

3. _____ Before the concrete arrives at a jobsite, the subgrade should be _____.
   A. compacted
   B. smoothed
   C. moistened
   D. All of the above.

4. _____ *True or False?* It is permissible to leave previously attached cement on forms when reusing the forms.

5. _____ Treating forms with _____ facilitates their removal after the concrete has hardened.
   A. grease
   B. oil or a scale remover
   C. oil or a releasing agent
   D. a releasing agent or water

6. _____ _____ are steel bars, square plates, or similar mechanisms installed in construction joints between concrete slabs.
   A. Dowels
   B. Rebars
   C. Studs
   D. Wales

7. _____ To reduce the effect of moisture transfer, a _____ is used between the subgrade and concrete slab.
   A. wood board
   B. moisture barrier
   C. piece of rigid insulation
   D. None of the above.

8. _____ When concrete is being placed, the concrete should be _____.
   A. placed as near as possible to its intended location
   B. worked over a long distance in the form
   C. placed in large quantities in one area
   D. All of the above.

9. _____ The separation of ingredients that can occur in areas where the grade slopes is called _____.
   A. segregation
   B. drifting
   C. separation
   D. Both A and C.

10. _____ *True or False?* Generally, concrete should be placed in horizontal layers having uniform thickness.

11. _____ Layers or lifts are usually _____ thick in reinforced concrete.
    A. 4″ to 10″
    B. 6″ to 12″
    C. 8″ to 14″
    D. 10″ to 16″

12. _____ Layers or lifts are usually up to _____ thick for nonreinforced applications.
    A. 12″
    B. 14″
    C. 16″
    D. 18″

13. _____ On large projects, the use of a _____ aids in moving concrete to isolated areas where other means can be almost impossible.
    A. ready-mix truck
    B. pump truck
    C. wheelbarrow
    D. track buggy

14. _____ What is the first step in placing concrete for slab construction?
    A. Excavate the area.
    B. Construct forms for concrete slab.
    C. Determine size and location of the slab.
    D. Determine the amount of concrete and type.

15. _____ Concrete placement in walls should begin at _____ and progress toward _____.
    A. either end, the center
    B. the bottom, the top
    C. the center, either end
    D. the right side, the left side

16. _____ Compacting the concrete is always necessary to be sure all spaces are filled and _____ are worked out.
    A. grooves
    B. lines
    C. air pockets
    D. None of the above.

17. _____ When placing fresh concrete on hardened concrete, the hardened concrete should be reasonably level, clean, _____, and _____.
    A. rough; moist
    B. smooth; dry
    C. smooth; moist
    D. dry; rough

18. _____ When concrete is placed on hardened concrete or rock, a layer of mortar _____ thick is placed on the hard surface to provide a cushion, prevent stone pockets, and secure a tight joint.
    A. 1/4″ to 1/2″
    B. 1/2″ to 1″
    C. 1″ to 1 1/2″
    D. 1 1/2″ to 2″

19. _____ *True or False?* Shotcrete is a mixture of Portland cement, aggregate, and water shot into place by means of compressed air.

20. _____ Which of the following statements about pneumatically applied concrete is *false*?
    A. Water content can be kept to a minimum.
    B. Pneumatic application works well where relatively thin sections and large areas are involved.
    C. The worker directing the nozzle controls the thickness of the concrete layer.
    D. In the wet mix process, ingredients are pumped through a hose to a nozzle where water is added.

Name _____

21. _____ The striking off of excess concrete to bring the top surface to the proper grade of elevation is called _____.
   A. screeding
   B. floating
   C. jointing
   D. edging

22. _____ *True or False?* When screeding, you should try to keep 1/2′ to 1′ of excess concrete in front of the screed as it is pulled forward.

23. _____ When concrete is placed against walls in an interior space, the concrete needs to be _____ in order to position it to the correct height and to level the surface.
   A. bull floated
   B. jointed or edged
   C. hand troweled or floated
   D. None of the above.

24. _____ Which of the following steps should be completed immediately following screeding and before bleed water has a chance to surface?
   A. Edging
   B. Floating
   C. Jointing
   D. Troweling

25. _____ *True or False?* Bleeding occurs more frequently with air-entrained concrete.

26. _____ Cracking due to shrinkage caused by drying or temperature change is controlled by _____.
   A. edging
   B. floating
   C. jointing
   D. troweling

27. _____ In sidewalk and driveway construction, the tooled joints are generally spaced at intervals equal to the width of the slab but not more than _____ apart.
   A. 1′
   B. 2′
   C. 5′
   D. 10′

28. _____ Which of the following is performed when a smooth, dense, hard surface is desired?
   A. troweling
   B. floating
   C. brooming
   D. All of the above.

29. _____ The process that produces a lightly roughened surface on surfaces that are often slippery when wet is called _____.
   A. troweling
   B. floating
   C. brooming
   D. curing

30. _____ *True or False?* Forms should be left in place at least until the concrete is strong enough to support its own weight and any other loads that may be immediately placed on it.

31. _____ If forms are tight and require wedging, only _____ should be used.
   A. grasp bars
   B. pinch bars
   C. steel wedges
   D. wooden wedges

32. _____ Air-entrained concrete _____.
    A. bleeds more than regular concrete
    B. requires less mixing water than regular concrete
    C. requires a wooden float for hand floating
    D. All of the above.

33. _____ During the early stage of concrete hardening, little or no _____ loss should be allowed.
    A. strength
    B. moisture
    C. hardness
    D. temperature

34. _____ A curing method that involves keeping approximately an inch of water on the concrete surface is called _____.
    A. membrane curing
    B. ponding
    C. pooling
    D. seeding

35. _____ According to the American Concrete Institute, cold weather is air temperature that averages less than _____ and is below 50°F more than half of each day for three weeks in a row.
    A. 0°F
    B. 30°F
    C. 40°F
    D. 45°F

36. _____ *True or False?* Concrete sets up more slowly, takes longer to finish, and gains strength more slowly in cold weather.

37. _____ If concrete freezes before it hardens, damage can reduce final compressive strength by as much as _____%.
    A. 33
    B. 50
    C. 75
    D. 90

38. _____ *True or False?* Concrete should never be placed on frozen subgrade.

39. _____ It is estimated that for every 10°F drop in concrete temperature, the set time _____.
    A. decreases by approximately one-third
    B. increases by approximately one-third
    C. increases by approximately one-half
    D. decreases by approximately one-half

40. _____ *True or False?* Air-entrained concrete is more susceptible to damage caused by freezing and thawing than normal concrete.

41. _____ In hot, dry, windy weather, concrete _____.
    A. sets more slowly
    B. is less susceptible to cracking
    C. loses its slump more quickly
    D. None of the above.

42. _____ All of the following methods can be used to prevent rapid drying of concrete in hot weather conditions, *except* _____.
    A. drying out the subgrade before placing concrete
    B. mixing the materials with chilled water or ice
    C. shading the operation
    D. using additives

Name _____

43. _____ The purpose of curing is to maintain conditions in the setting concrete that encourage complete _____.
    A. hardness
    B. strength
    C. hydration
    D. density

44. _____ Joints that separate different parts of a structure to permit both vertical and horizontal movements are called _____ joints.
    A. construction
    B. control
    C. isolation
    D. segregation

45. _____ _____ joints provide for movement in the same plane as the slab or wall is positioned and compensate for contraction caused by drying shrinkage.
    A. Isolation
    B. Construction
    C. Contraction
    D. Control

46. _____ _____ joints provide for no movement across the joint and are only stopping places in the process of casting.
    A. Control
    B. Construction
    C. Isolation
    D. Building

47. _____ When coloring concrete, the amount of pigment should never exceed _____% of the weight of the cement.
    A. 10
    B. 25
    C. 40
    D. 50

48. _____ *True or False?* The only difference between the one-course and two-course methods is that the two-course method uses a base coat of conventional concrete.

49. _____ One of the most common methods used to produce exposed aggregate finishes is the _____ method.
    A. brooming
    B. terrazzo
    C. seeding
    D. None of the above.

50. _____ Placing a thin topping course of concrete containing special aggregates over a base of regular concrete is a technique used for _____ construction.
    A. form
    B. terrazzo
    C. aggregate
    D. stone

51. _____ The effect produced by applying a dash coat of mortar over freshly leveled concrete is called a(n) _____ finish.
    A. marble
    B. mineral
    C. travertine
    D. ashlar

52. _____ When is the best time for scoring concrete?
    A. While it is curing.
    B. After darbying or bull floating.
    C. After hand floating.
    D. After screeding.

53. _____ Patterns can be created using _____ strips of wood, plastic, metal, or masonry units.
    A. scoring
    B. grooving
    C. divider
    D. stamping

54. _____ Stamping of concrete usually begins after screeding and _____ have been completed.
    A. coloring
    B. finishing
    C. jointing
    D. curing

55. _____ *True or False?* Aluminum oxide grains are sparkling black in color and are used to make "sparkling concrete."

56. _____ Cast-in-place (CIP) concrete walls are made with _____ concrete that is transported to the jobsite and placed into forms that are erected on the site.
    A. sparkling
    B. flat
    C. reused
    D. ready-mix

57. Vertical fins or _____ can produce shadow effects on a concrete wall.

_____

58. _____ *True or False?* Joint patterns and tie holes can be used in a repeating pattern for architectural effect.

59. _____ One of the best methods of obtaining color in a cast-in-place concrete wall is through the use of _____.
    A. exposed aggregate
    B. admixtures
    C. fine aggregate
    D. ribs

60. _____ In cast-in-place window walls, _____ Portland cement concrete is well suited because the color is permanent.
    A. black
    B. brown
    C. gray
    D. white

61. _____ *True or False?* Pan joist construction is a two-way reinforced structural system using a ribbed slab formed with pans.

62. _____ A flat plate roof and floor system _____.
    A. requires supporting columns in a straight line
    B. is a two-way reinforced concrete framing system
    C. has slabs that provide spans up to 40′
    D. All of the above.

63. _____ What roof and floor system is shown in the following image?
    A. Flat plate
    B. Waffle
    C. Flat slab
    D. Pan joist

Goodheart-Willcox Publisher

Name _____

64. _____ Which of the following types of cast-in-place concrete roof and floor systems is designed for heavy roof loads with large, open bays below?
    A. Flat plate
    B. Flat slab
    C. Pan joist
    D. Waffle

65. _____ _____ construction is especially suited for buildings greater than 10,000 sq ft with 20′ or higher side walls that incorporate repetition in panel size and appearance.
    A. Cast-in-place
    B. Tilt-up
    C. Panel
    D. Both A and B.

66. _____ *True or False?* Spread footings are used for most tilt-up buildings.

67. _____ Tilt-up panels _____.
    A. can have a patterned or textured surface
    B. do *not* require reinforcing steel
    C. have bracing that is installed once the panels are in vertical position
    D. All of the above.

68. _____ The most widely used prestressed, precast units are _____-tees.
    A. single
    B. double
    C. hollow-core
    D. waffle

69. _____ The type of prestressed, precast unit that is generally used where long spans, beginning at about 30′, are needed is the _____ unit.
    A. double-tee
    B. hollow-core
    C. single-tee
    D. waffle

70. _____ Which of the following prestressed, precast units provide a flush ceiling with minimum depth required for the roof or floor system?
    A. Double-tees
    B. Hollow-cored
    C. Single-tees
    D. Solid-cored

71. _____ *True or False?* Precast concrete window walls may be cast as curtain walls or loadbearing walls.

Notes

Name _____ Date _____ Class _____

JOB 1: Identifying Common Masonry Materials

OBJECTIVE: After completing this job, you will be able to identify and classify common masonry materials.

TEXTBOOK REFERENCE: Study the appropriate sections in Chapter 7, *Clay Masonry Materials*; Chapter 8, *Concrete Masonry Units*; and Chapter 9, *Stone* before starting this job.

Equipment

To complete this job, you will need the following tools and materials:

- Pencil
- Flexible measuring tape or folding rule
- Textbook for reference
- 16 common masonry materials (selected by instructor). Each material should be identified with a letter (A–P).

Recommended Procedure

1. Place each masonry material into one of the following groups: Completed ☐
   - Clay brick
   - Clay tile
   - Concrete masonry unit
   - Sand lime brick
   - Glass block
   - Natural stone
   - Manufactured stone

2. Write the accepted name and category (see list in No. 1) of each masonry unit in the spaces below. Use your rule or tape to measure units to help you identify the materials. Refer to your text for sizes and names. Completed ☐

   A. _____
   B. _____
   C. _____
   D. _____
   E. _____
   F. _____
   G. _____
   H. _____
   I. _____
   J. _____
   K. _____
   L. _____
   M. _____

N. _____

O. _____

P. _____

Instructor's Initials: _____

Date: _____

Job 1 Review

After completing this job successfully, identify the following masonry materials.

1. Category: _____

   Name: _____

   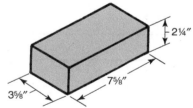

2. Category: _____

   Name: _____

   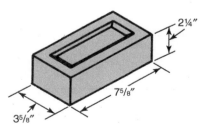

3. Category: _____

   Name: _____

4. Category: _____

   Name: _____

Name _____

5. Category: _____

   Name: _____

   6" × 12"   8" × 12"

   Goodheart-Willcox Publisher

6. Category: _____

   Name: _____

   Goodheart-Willcox Publisher

7. Category: _____

   Name: _____

   Goodheart-Willcox Publisher

8. Category: _____

   Name: _____

   Goodheart-Willcox Publisher

9. Category: _____

   Name: _____

   Goodheart-Willcox Publisher

Job 1 Identifying Common Masonry Materials

10. Category: _____

　　Name: _____

Goodheart-Willcox Publisher

　　　　　　　　　Score: _____

Name _____ Date _____ Class _____

JOB 2: Identifying Types of Clay Brick

OBJECTIVE: After completing this job, you will be able to identify various types of clay brick.

TEXTBOOK REFERENCE: Study the appropriate section in Chapter 7, *Clay Masonry Materials,* before starting this job.

Equipment

To complete this job, you will need the following tools and materials:

- Pencil
- Flexible measuring tape or folding rule
- Textbook for reference
- 16 common types of clay brick (selected by instructor). Each type of brick should be identified with a letter (A–P).

Recommended Procedure

1. Place each clay masonry unit into one of the following groups: Completed ☐
- Building brick
- Facing brick
- Hollow brick
- Paving brick
- Ceramic glazed brick
- Thin brick veneer units
- Sewer or manhole brick

2. Write the accepted name and category (see list in No. 1) of each clay masonry unit in the   Completed ☐
spaces below. Use your rule or tape to measure the brick to help you identify the units. Refer
to your text for sizes and names.

A. _____

B. _____

C. _____

D. _____

E. _____

F. _____

G. _____

H. _____

I. _____

J. _____

K. _____

L. _____

M. _____

N. _____

O. _____

P. _____

Instructor's Initials: _____

Date: _____

Job 2 Review

After completing this job successfully, identify the following clay masonry materials. Use these categories—building brick, facing brick, hollow brick, paving brick, ceramic glazed brick, thin brick veneer, and sewer or manhole brick.

1. _____

2. _____

3. _____

Goodheart-Willcox Publisher

Name _____

4. _____

5. _____

6. _____

7. _____

Job 2　Identifying Types of Clay Brick

8. Write the category of each masonry unit next to its name on the lines provided in No. 1. The categories are as follows:
   - Clay brick
   - Clay tile
   - Concrete masonry unit
   - Sand lime brick
   - Glass block
   - Natural stone
   - Manufactured stone

Score: _____

Name _____ Date _____ Class _____

JOB 3: Arranging Brick in the Five Basic Structural Bonds

OBJECTIVE: After completing this job, you will be able to arrange bricks in each of the five basic structural bonds—running, common, Flemish, English, and stack.

TEXTBOOK REFERENCE: Study the appropriate section in Chapter 7, *Clay Masonry Materials,* before starting this job.

Equipment

To complete this job, you will need the following tools and materials:

- 45 common bricks
- 36 half bricks (snap headers)
- 8 three-quarter closures
- 8 quarter brick closures

Recommended Procedure

1. Arrange the bricks in a single wythe (dry bond) in running bond nine courses high. See **Figure 3-1**. Completed ☐

> **WARNING!**
> Lay the bricks on a flat smooth surface as though the wall were horizontal rather than vertical. Nine courses of bricks might fall over and injure someone.

2. Arrange the bricks in a single wythe (dry bond) in common bond with 6th course headers. Lay the bricks nine courses high on a flat surface as in Step 1. Headers should be used for the 2nd and 8th courses. Completed ☐

3. Arrange the bricks in a single wythe (dry bond) in common bond with 6th course Flemish headers. Lay the bricks nine courses high. Completed ☐

4. Arrange the bricks in a single wythe (dry bond) in Flemish bond. Lay the bricks nine courses high. Completed ☐

5. Arrange the bricks in a single wythe (dry bond) in English bond. Lay the bricks nine courses high. Completed ☐

6. Arrange the bricks in a single wythe (dry bond) in stack bond. Lay the bricks nine courses high. Completed ☐

Instructor's Initials: _____

Date: _____

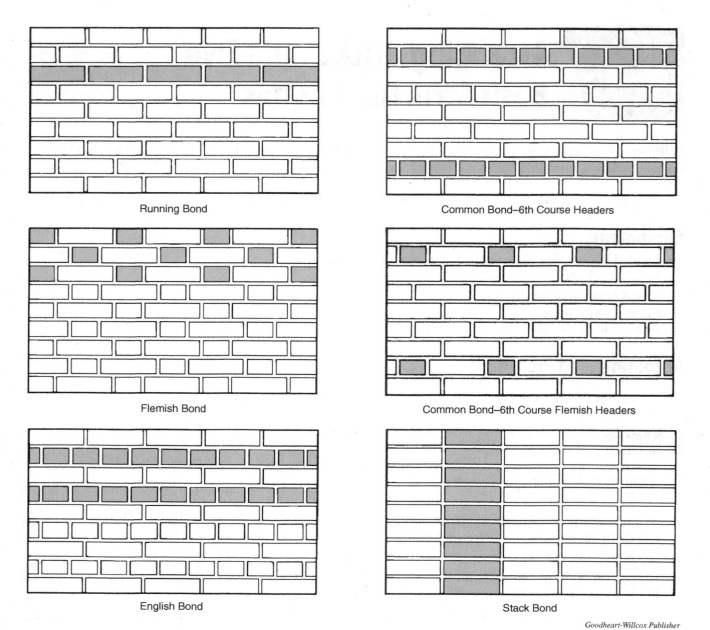

Figure 3-1. The basic structural bonds.

Name _____

Job 3 Review

After completing this job successfully, identify the following structural bonds. The types are running, common, Flemish, English, and stack.

1. _____

2. _____

3. _____

4. _____

Job 3 Arranging Brick in the Five Basic Structural Bonds

5. _____

Goodheart-Willcox Publisher

Score: _____

Name _____ Date _____ Class _____

JOB 4: Identifying Common Concrete Masonry Units

OBJECTIVE: After completing this job, you will be able to identify and classify common concrete masonry units.

TEXTBOOK REFERENCE: Study the appropriate section in Chapter 8, *Concrete Masonry Units*, before starting this job.

Equipment
To complete this job, you will need the following tools and materials:
- Pencil
- Flexible tape or folding rule
- Textbook for reference
- 16 common types of concrete masonry units (selected by instructor)

Recommended Procedure

1. Place each concrete masonry unit into one of the following categories: Completed ☐
   - Concrete brick
   - Loadbearing concrete block
   - Nonloadbearing concrete block
   - Calcium silicate face brick
   - Prefaced concrete units
   - Units for catch basins and manholes

2. Write the accepted name and category (see list in No. 1) of each concrete masonry unit in the spaces below. Use your rule to measure the units to help you determine their identification. Refer to your text for sizes and names.   Completed ☐

   A. _____
   B. _____
   C. _____
   D. _____
   E. _____
   F. _____
   G. _____
   H. _____
   I. _____
   J. _____
   K. _____
   L. _____
   M. _____
   N. _____

O. _____

P. _____

Instructor's Initials: _____

Date: _____

Job 4 Review

After completing this job successfully, identify the following concrete masonry units. The possible types are concrete brick, loadbearing concrete block, nonloadbearing concrete block, calcium silicate face brick, prefaced concrete units, and units for catch basins and manholes.

1. _____

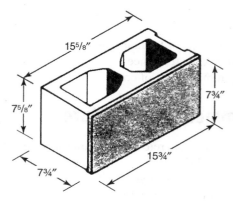

Goodheart-Willcox Publisher

2. _____

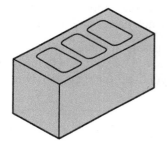

Goodheart-Willcox Publisher

3. _____

Air Vol Block

Name _____

4. _____

5. _____

Score: _____

Job 4   Identifying Common Concrete Masonry Units   121

Notes

Name _____ Date _____ Class _____

JOB 5 — Identifying Common Concrete Block Shapes by Name

OBJECTIVE: After completing this job, you will be able to identify common concrete block shapes by name.

TEXTBOOK REFERENCE: Study the appropriate section in Chapter 8, *Concrete Masonry Units,* before starting this job.

Equipment

To complete this job, you will need the following tools and materials:

- Pencil
- Flexible tape or folding rule
- Textbook for reference
- 12 common types of concrete masonry units (selected by instructor). Each unit should be identified with a letter.

Recommended Procedure

1. Examine each concrete masonry unit for characteristics that help to identify its type (name). For example, does it have cells? What are its dimensions? Does it have a unique shape? Where might it be used in construction? How much does it weigh?   Completed ☐

2. Record the name of each concrete masonry unit beside its identification letter in the following spaces. Examples: jamb block, corner block, stretcher. See the list of names in **Figure 5-1.**   Completed ☐

A. _____

B. _____

C. _____

D. _____

E. _____

F. _____

G. _____

H. _____

I. _____

J. _____

K. _____

L. _____

Copyright Goodheart-Willcox Co., Inc.
May not be reproduced or posted to a publicly accessible website.

Stretcher block	Lintel block	Depressed-face unit
Corner block	Solid block	Screen block
Double corner block	Slab or partition block	Preface unit
Sash block	Half-hi full length block	Lightweight stretcher
Sash half block	Half-hi half length block	Sound block
Jamb block		Slump block
Jamb half block	Soffit block	Ground face unit
Single bull nose block	Insulating block	Control joint block
Double bull nose block	Ribbed unit	Solid top block
Double bull nose face block	Fluted unit	Channel block
	Stri-face unit	One-piece chimney block
1/4 block	Split-fluted unit	
3/4 block	Split-face unit	Pilaster block
Full cut out header block	Sculptured-face unit	Standard wall two-core block
	Offset-face unit	
Half cut out header block		Glazed block

Figure 5-1. Names of concrete masonry units.

Goodheart-Willcox Publisher

Instructor's Initials: _____

Date: _____

Job 5 Review

After completing this job successfully, identify the following common concrete block shapes by name.

1. _____

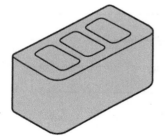

Goodheart-Willcox Publisher

2. _____

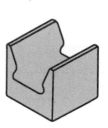
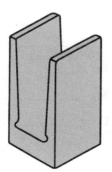

Goodheart-Willcox Publisher

Name ______________________________

3. ______________________________

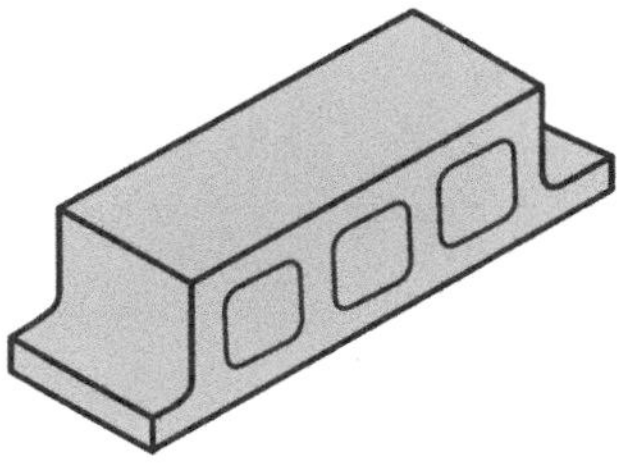

Goodheart-Willcox Publisher

4. ______________________________

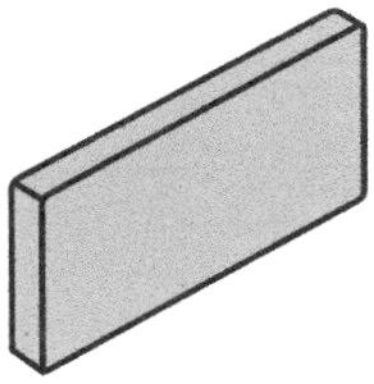

Goodheart-Willcox Publisher

5. ______________________________

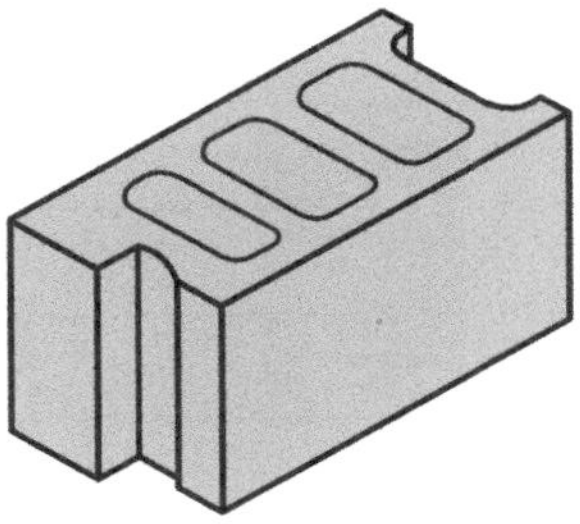

Goodheart-Willcox Publisher

6. ______________________________

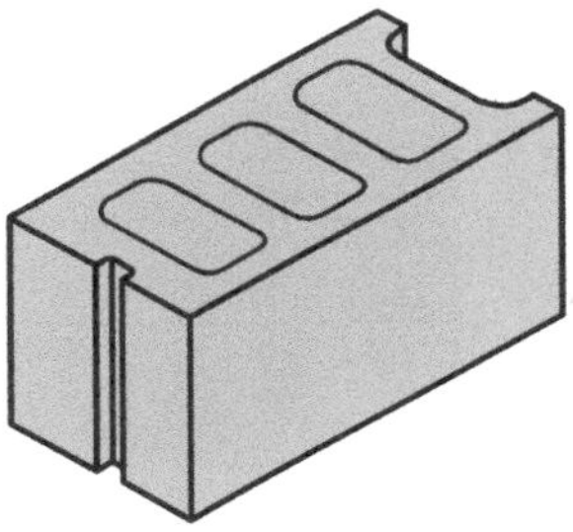

Goodheart-Willcox Publisher

7. ______________________________

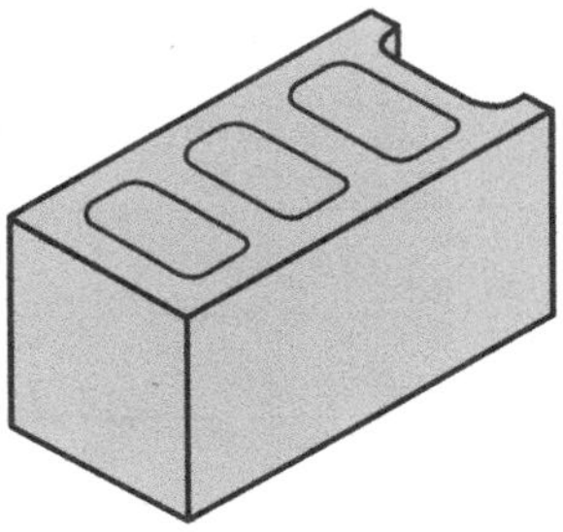

Goodheart-Willcox Publisher

8. ______________________________

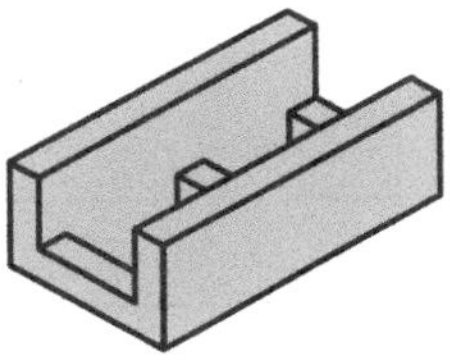

Goodheart-Willcox Publisher

Score: ______________________________

Notes

Name _____ Date _____ Class _____

Identifying Common Building Stone Samples

OBJECTIVE: After completing this job, you will be able to identify and classify common building stones.

TEXTBOOK REFERENCE: Study the appropriate section in Chapter 9, *Stone,* before starting this job.

Equipment

To complete this job, you will need the following tools and materials:

- Pencil
- Textbook for reference
- 10 common types of building stone (selected by instructor). Each sample will be identified with a letter.

Recommended Procedure

1. Classify each sample as igneous (volcanic), sedimentary, or metamorphic. Write the classification beside its identification letter in the space provided following Step 2.   Completed ☐

2. Identify each sample and record the name beside its identification letter and group name below. Refer to the text for help in identifying the samples. The following stone types are typical: granite, traprock, sandstone, limestone, marble, slate, schist, gneiss, quartzite.   Completed ☐

A. _____

B. _____

C. _____

D. _____

E. _____

F. _____

G. _____

H. _____

I. _____

J. _____

Instructor's Initials: _____

Date: _____

Job 6 Review

After completing this job successfully, identify the following types of building stone.

1. _____

Goodheart-Willcox Publisher

2. _____

Goodheart-Willcox Publisher

3. _____

pics721/Shutterstock.com

Name _____

4. _____

Goodheart-Willcox Publisher

Score: _____

Notes

Name _____ Date _____ Class _____

Measuring Mortar Materials and Mixing Mortar

OBJECTIVE: After completing this job, you will be able to measure mortar materials using recommended practices and mix mortar to the proper consistency.

TEXTBOOK REFERENCE: Study the appropriate section in Chapter 10, *Mortar and Grout*, before starting this job.

Equipment

To complete this job, you will need the following tools and materials:
- Bag of masonry cement or Type I Portland cement and Type S hydrated lime
- Masonry sand
- Water suitable for mixing mortar
- One cubic foot measuring box
- Bucket or pail for measuring water
- Mixing box
- Mortar hoe
- Eye and hand protection
- Mason's trowel

Recommended Procedure

1. Collect the materials and tools needed to measure and mix mortar materials. Arrange the items so they are convenient to the work area.   Completed ☐

WARNING!
Wear eye and hand protection during the following procedure.

2. Add 1 cu ft of sand to the mixing box. Use the one cubic foot measuring box to measure the sand.   Completed ☐

3. Add 1/3 cu ft of masonry cement to the mixing box. Use the one cubic foot measuring box to measure the masonry cement. Note: If Portland cement and lime are used instead of masonry cement, use 1/3 cu ft of cement and 1/12 cu ft of lime.   Completed ☐

4. Mix the sand and cement together with the mortar hoe. These ingredients must be mixed thoroughly before adding water. Work the materials from one end of the box to the other several times.   Completed ☐

5. Add 2/3 to 3/4 of the water and mix with the mortar hoe until the batch is uniformly wet. Experience will help you determine how much water to add. (There are 8.33 lbs per gallon of water.) The general guideline is to add the maximum amount of water consistent with workability to provide maximum tensile bond strength. Your instructor will help you estimate how much water will be needed for this batch.   Completed ☐

6. Add more water to the batch and mix until the ingredients are thoroughly mixed and the batch is smooth and creamy. Check to see if the mortar has the right plasticity and adhesiveness by cutting ridges with the trowel or spade. Well-mixed mortar forms sharp ridges.   Completed ☐

7. Dispose of the mortar as directed by your instructor. Clean all tools and return all unused materials to their proper storage place.   Completed ☐

Instructor's Initials: _____

Date: _____

Job 7 Review

After completing the job successfully, answer the following questions.

1. Since mortar can be mixed using a variety of mix proportions, why is it important that the ingredients be measured exactly the same each time?

   _____
   _____
   _____

2. Why not add all of the water to the mix instead of adding about two-thirds to three-fourths of it after the cement and sand have been mixed?

   _____
   _____
   _____

3. How do you know when the mortar is mixed to the right consistency?

   _____
   _____
   _____

4. Why will mortar stick to the trowel, but concrete will not?

   _____
   _____
   _____

5. What is the difference between masonry cement and Portland cement?

   _____
   _____
   _____

Score: _____

Name _____ Date _____ Class _____

JOB 8: Identifying Anchors, Ties, and Joint Reinforcement

OBJECTIVE: After completing this job, you will be able to identify anchors, ties, and joint reinforcement commonly used in concrete and masonry construction.

TEXTBOOK REFERENCE: Study the appropriate section in Chapter 11, *Anchors, Ties, and Reinforcement,* before starting this job.

Equipment

To complete this job, you will need the following tools and materials:

- Pencil
- Textbook for reference
- 12 common anchors, ties, and joint reinforcement (selected by instructor) used in concrete and masonry construction. Each item will be identified with a number.

Recommended Procedure

1. Identify each sample as an anchor, tie, or joint reinforcement. Write the sample type on the lines in Step 2.   Completed ☐

2. Identify the name of each sample and record the name beside its type on the lines provided. Refer to the text for help in identifying each sample. The following names are typical: Completed ☐

Rectangular ties	Epoxy adhesive bolts
Ladder type reinforcement	Reinforcing bar
Z ties	Dovetail anchor
Truss type reinforcement	"J" anchor bolt
Corrugated ties	"L" anchor bolt
Tab type reinforcement	Anchor strap
Adjustable unit ties	Hardware cloth
Expansion bolts	Through rods
Adjustable assemblies	Power-driven pin

A. _____ G. _____

B. _____ H. _____

C. _____ I. _____

D. _____ J. _____

E. _____ K. _____

F. _____ L. _____

Instructor's Initials: _____

Date: _____

Job 8 Review

After completing this job successfully, answer the following questions.

1. Why are anchor bolts used in masonry construction?

2. Why does the embedded end of most anchor bolts have a head, plate, or bent angle?

3. How is joint reinforcement used in brick or block masonry?

4. In what type of application (wall) are corrugated fasteners generally used?

5. How does a band steel anchor bolt develop its holding power?

6. When might an adjustable assembly be used?

Score: _____

Name _____ Date _____ Class _____

JOB 9 — Loading the Trowel and Spreading a Mortar Bed for Brick and Block

OBJECTIVE: After completing this job, you will be able to spread a mortar bed for clay brick and concrete block using proper technique.

TEXTBOOK REFERENCE: Study the appropriate section in Chapter 4, *Safety*; Chapter 12, *Laying Brick*; and Chapter 13, *Laying Block* before starting this job.

Equipment

To complete this job, you will need the following tools and materials:
- Standard mason's trowel
- 4 clay bricks
- 2 concrete blocks
- Mortar
- Mortar board
- Eye and hand protection

Recommended Procedure

1. Review safety precautions to be taken when using cementitious materials.   Completed ☐
2. Collect the tools and materials needed for this job and arrange a work area that is convenient and functional.   Completed ☐
3. Mix a batch of mortar following the procedure described in Job 7. Place a generous amount of mortar on your mortar board and work it into a neat pile.   Completed ☐
4. Practice loading the trowel by following these steps or another procedure that feels natural to you: Completed ☐
   A. Grasp the trowel in your dominant hand, with the thumb on top of the ferrule and the fingers under the handle.
   B. Work the mortar into a pile into the center of the mortar board.
   C. Smooth off a place with a backhand stroke.
   D. Cut a small amount from the larger pile with a pulling action.
   E. Scoop up the small pile with a quick movement of the trowel. **Figure 9-1** shows the loaded trowel.

Author's image taken at Job Corps, Denison, IA

Figure 9-1. Trowel loaded and ready.

Copyright Goodheart-Willcox Co., Inc.
May not be reproduced or posted to a publicly accessible website.

5. Lay down a mortar bed for a course of bricks using a quick turn of your wrist toward your body and a backward movement of your arm. As the trowel is nearly empty, tip the trowel blade even more to help the remaining mortar slide off.   Completed ☐

6. Furrow the mortar bed with the point of the trowel. This helps to form a uniform bed for solid masonry units. See **Figure 9-2**.   Completed ☐

Author's image taken at Job Corps, Denison, IA

Figure 9-2. Furrowing the first course to form a uniform bed.

7. Press several bricks into the bed of mortar. Do not apply head joints, since that procedure has not yet been discussed. Note the pressure needed to set the bricks.   Completed ☐

8. Practice the procedure for loading the trowel and spreading a bed of mortar for bricks until it feels comfortable and you can judge the right amount of mortar needed.   Completed ☐

9. Repeat the process described above, but this time prepare a bed for concrete blocks. The same approach is used, but a wider bed is needed.   Completed ☐

10. Using both hands, grasp two cell webs and place a concrete block on the mortar bed, pressing it lightly into place. The thicker edge of the shell is on top to provide a wider mortar bed for the next course. This process will be practiced in more detail later.   Completed ☐

11. Practice laying down a mortar bed for concrete blocks until you are confident of the process.   Completed ☐

12. Clean the mortar from the masonry units, mortar board, work surface, and tools. Return tools and materials to their proper places. Dispose of the mortar as directed by your instructor.   Completed ☐

WARNING!
Contact with wet (plastic) concrete, cement, mortar, grout, or cement mixtures can cause skin irritation, severe chemical burns, or serious eye damage. Wear waterproof gloves, a long-sleeved shirt, full-length trousers, and proper eye protection when working with these materials. Wash wet mortar from your skin immediately. Flush your eyes with clear water immediately upon contact. Seek medical attention if you experience a reaction to contact with these materials.

Instructor's Initials: _____

Date: _____

Name _____

Job 9 Review

After completing this job successfully, answer the following questions.

1. Construction is a dangerous business; therefore, safety is a major consideration on the job. Describe a properly-dressed worker in the masonry trades.

   _____
   _____
   _____
   _____

2. Describe the recommended method of holding a mason's trowel.

   _____
   _____
   _____

3. What did you find to be the most difficult part of loading the trowel and spreading a bed of mortar?

   _____
   _____
   _____

4. Knowing when and just how much to temper (add water to) mortar is an important skill that every mason must learn. Did you need to temper your mortar? How did you know that it needed more water?

   _____
   _____
   _____
   _____

5. How long do you generally have to use up a batch of mortar before it has to be discarded?

   _____
   _____

   Score: _____

Notes

Name _____ Date _____ Class _____

JOB 10: Forming a Head Joint on Brick and Block

OBJECTIVE: After completing this job, you will be able to form a head joint on brick and block using the proper technique.

TEXTBOOK REFERENCE: Study the appropriate sections in Chapter 4, *Safety*; Chapter 12, *Laying Brick*; and Chapter 13, *Laying Block* before starting this job.

Equipment

To complete this job, you will need the following tools and materials:

- Standard mason's trowel
- 6 clay bricks
- 4 concrete blocks
- Mortar
- Mortar board
- Eye and hand protection

Recommended Procedure

1. Review safety precautions for using cementitious materials.   Completed ☐
2. Collect the tools and materials needed for this job and arrange a work area that is convenient and functional.   Completed ☐
3. Mix a batch of mortar, following the procedure described in Job 7. Place a generous amount of mortar on your mortar board and work it into a neat pile.   Completed ☐
4. Lay down a mortar bed for a course six bricks long. Furrow the mortar bed so it is ready to receive the base course of bricks. Place the first brick on the bed without a head joint.   Completed ☐
5. Pick up a brick in your nondominant hand, holding the brick about midpoint across it. Load your trowel with the other hand. Apply mortar to the end of the brick with a swiping or throwing action across the end of the brick. This motion should attach the mortar to the brick and form it into a wedge shape. See **Figure 10-1**.   Completed ☐

Author's image taken at Job Corps, Denison, IA

Figure 10-1. Forming a head joint on a brick.

6. Bevel the mortar slightly at the corners with the trowel. Press the brick into the mortar bed and against the brick already in place. Some mortar should squeeze out between the bricks. Completed ☐

7. Repeat the process several times to develop the proper hand/wrist action. The purpose of this job is to learn how to apply head joints, not lay brick. Completed ☐

8. Remove the brick and mortar. Clean the mortar off the brick and work the surface for the next operation. Completed ☐

9. Lay down a mortar bed for a course of concrete block using the procedure practiced in Job 9. Completed ☐

10. Place one concrete block on the mortar bed and press it into place. Do not apply a head joint to the block. Completed ☐

11. Set several blocks on end so that the ears face up to receive the mortar. Scoop some mortar on the trowel, but do not fully load it. Use a downward swiping action to apply the mortar to the ears of the blocks. Completed ☐

12. Next, press down the mortar on the inside of each ear of the block to attach it to the block. If the mortar is not attached sufficiently, it will fall off when the block is lifted and turned for placement in the wall. Completed ☐

13. Lift the block with both hands and place it on the mortar bed next to the first block. Press down and against the other block to form a good watertight joint. Mortar should squeeze out. Completed ☐

14. Practice applying head joints to blocks several times to develop the proper action and to apply the right amount of mortar to the block. Completed ☐

15. Clean the mortar from the masonry units, mortar board, work surface, and tools. Return the tools and materials to their proper places and dispose of the mortar as directed by your instructor. Completed ☐

Instructor's Initials: _____

Date: _____

Job 10 Review

After completing this job successfully, answer the following questions.

1. One of the problems that student masons have is keeping mortar on the brick or block as they move it toward its destination. What do you think might be the cause for this?

   _____
   _____
   _____

2. When forming head joints, how did you know that you had applied enough mortar to form a full head joint?

   _____
   _____

3. If a little mortar is good, then why not apply lots of mortar to form the head joint?

   _____
   _____
   _____

Name _____

4. Did you load your trowel with more or less mortar when forming head joints on block than when forming head joints on brick?

   _____

   _____

5. If some of the mortar falls off when placing a brick or block, should you place the brick or block and then fill the joint?

   _____

   _____

Score: _____

Notes

Name _____ Date _____ Class _____

JOB 11: Cutting Brick with a Brick Hammer

OBJECTIVE: After completing this job, you will be able to cut brick with a brick hammer using the proper technique.

TEXTBOOK REFERENCE: Study the appropriate section in Chapter 12, *Laying Brick*, before starting this job.

Equipment

To complete this job, you will need the following tools and materials:

- Brick hammer
- Several bricks
- Eye and hand protection

Recommended Procedure

1. A brick hammer is most frequently used by masons to cut brick. Begin by holding the brick in one hand, keeping the fingers away from the side where the cutting will take place. With the other hand, strike the brick with light blows with the blade edge of the brick hammer along the line where the cut is planned. Completed ☐

2. Turn the brick over to the adjacent edge and continue the cut along that side. Repeat the process until all sides have been scored. Completed ☐

WARNING!
Be sure you are wearing safety glasses during the following procedure.

3. Strike the face of the brick with a sharp blow of the hammer. The brick should break along the scored lines. Completed ☐

4. Repeat the procedure several times to develop the proper technique. Completed ☐

Instructor's Initials: _____

Date: _____

Job 11 Review

After completing this job successfully, answer the following questions.

1. Masons most often use a brick hammer to cut brick. Why do you think this is true?

_____
_____
_____

2. As a novice brick mason, why do you think that it is a good idea to draw a line on the brick where you wish to break it with the brick hammer?

_____

_____

_____

3. Why do you need to cut a groove along all sides of the brick before striking the brick to break it?

_____

_____

_____

4. When you have scored the brick on all sides, which part of the brick hammer do you use to strike the brick?

_____

_____

5. What is the most important piece of safety equipment to wear when cutting brick with a brick hammer?

_____

_____

Score: _____

Name _____ Date _____ Class _____

JOB 12: Cutting Brick with a Brick Set Chisel

OBJECTIVE: After completing this job, you will be able to cut brick with a brick set chisel using the proper technique.

TEXTBOOK REFERENCE: Study the appropriate section in Chapter 12, *Laying Brick*, before starting this job.

Equipment

To complete this job, you will need the following tools and materials:

- Pencil
- Brick hammer
- Brick set chisel
- Several bricks
- Eye and hand protection

Recommended Procedure

1. When a more accurate, straight cut is needed, a brick set chisel may be used to cut brick. First, mark the brick with a pencil or other marking device where the cut is to be made.   Completed ☐

2. Place the brick on a soft surface, such as soil or a board. Hold the chisel end of the brick set vertically with the flat side of the blade facing the direction of the finished cut.   Completed ☐

WARNING!
Be sure to wear safety glasses during the following procedure.

3. Strike the brick set sharply with the brick hammer. The resulting cut should be relatively smooth and not require additional chipping before use in the wall.   Completed ☐

4. Repeat the procedure several times to develop the proper technique.   Completed ☐

Instructor's Initials: _____

Date: _____

Job 12 Review

After completing this job successfully, answer the following questions.

1. Why would you use a brick set chisel rather than a brick hammer to cut a brick?

   _____
   _____
   _____

2. Why shouldn't you place brick on a hard surface, such as a slab of concrete, to cut it with a brick set chisel?

   _____
   _____
   _____

3. Which side of the chisel should be facing the direction of the finished cut and why?

   _____

   _____

   _____

4. Why is it necessary to keep the brick set chisel sharp?

   _____

   _____

   _____

                              Score: _____

Name _____ Date _____ Class _____

JOB 13: Cutting Brick with a Mason's Trowel

OBJECTIVE: After completing this job, you will be able to cut brick with a brick mason's trowel using the proper technique.

TEXTBOOK REFERENCE: Study the appropriate section in Chapter 12, *Laying Brick,* before starting this job.

Equipment

To complete this job, you will need the following tools and materials:
- Standard mason's trowel
- Several bricks
- Eye and hand protection

Recommended Procedure

1. A mason's trowel can be used to cut softer brick but is not recommended for most cutting, especially hard brick. When cutting a brick with a trowel, hold the brick in one hand, keeping your fingers away from the side where the cutting is to be done. Completed ☐

WARNING!
Be sure you are wearing safety glasses during the following procedure.

2. Hold the brick down away from your face. Strike the brick with the edge of the trowel using a quick, sharp blow at the spot where the cut (break) is intended. Completed ☐

3. Repeat the process several times to develop the technique. Completed ☐

4. Clean up the pieces and return all tools and materials to their proper place. Completed ☐

Instructor's Initials: _____

Date: _____

Job 13 Review

After completing this job successfully, answer the following questions.

1. Why shouldn't a mason's trowel be used to cut hard brick?

   _____
   _____
   _____

2. Why do some masons cut hard brick with a trowel?

   _____
   _____
   _____

3. Why is cutting brick with a trowel basically unsafe?

_____

_____

_____

4. What type of masonry unit can be cut with a brick mason's trowel?

_____

_____

Score: _____

Name _____ Date _____ Class _____

JOB 14: Cutting Brick with a Masonry Saw

OBJECTIVE: After completing this job, you will be able to cut bricks with a masonry saw using the proper technique.

TEXTBOOK REFERENCE: Study the appropriate section in Chapter 12, *Laying Brick*, before starting this job.

Equipment

To complete this job, you will need the following tools and materials:
- Power masonry saw
- Several bricks
- Eye protection, face shield, and ear protection

Recommended Procedure

1. When an exact, smooth cut is required, a power masonry saw may be used. This process is slower but produces the highest-quality cut. Ask your instructor to demonstrate the proper and safe use of the masonry saw. — Completed ☐

WARNING!
A power masonry saw is a dangerous piece of equipment to operate. Safety glasses or goggles are required. A face shield should also be worn to protect from flying chips. Ear protection is needed because of the amount of noise created by sawing.

2. Hold the brick firmly against the fence and move the saw slowly through the cut. Do not rush the cut or let the brick move while making the cut. — Completed ☐
3. Move the saw back to its original position before removing the pieces. — Completed ☐
4. Repeat the process several times to develop confidence in safely operating a masonry saw. — Completed ☐

TRADE TIP
Use the same procedure to cut concrete blocks using a masonry saw.

Instructor's Initials: _____

Date: _____

Job 14 Review

After completing this job successfully, answer the following questions.

1. When would you choose to use a masonry saw to cut brick rather than a brick hammer, trowel, or brick set chisel?

_____
_____
_____

2. How does the blade on a masonry saw cut brick?

_____
_____
_____

3. What safety equipment is needed when operating a masonry saw?

_____
_____

4. How could you tell if you were moving the saw through the cut too quickly?

_____
_____
_____

5. What should you look for in a routine examination of a masonry saw before using it?

_____
_____
_____

Score: _____

Name _____ Date _____ Class _____

JOB 15: Cutting Concrete Block with a Brick Hammer and Blocking Chisel

OBJECTIVE: After completing this job, you will be able to cut concrete blocks with a brick hammer and blocking chisel using the proper technique.

TEXTBOOK REFERENCE: Study the appropriate section in Chapter 13, *Laying Block,* before starting this job.

Equipment

To complete this job, you will need the following tools and materials:

- Brick hammer
- Blocking chisel
- Several concrete blocks
- Eye and hand protection

Recommended Procedure

1. Even though concrete blocks are available in half-length units as well as full-length units, it is sometimes necessary to cut blocks to fit. When using the blocking chisel, hold the beveled edge toward you so the piece to be cut off is facing away.　　　Completed ☐

> **WARNING!**
> Be sure you are wearing safety glasses during the following procedure.

2. Place the chisel on the line where the cut is to be made. Strike the chisel with the brick hammer to score a line where the cut is to be made. See **Figure 15-1**.　　　Completed ☐

Author's image taken at Job Corps, Denison, IA

Figure 15-1. Cutting a concrete block with a blocking chisel and mason's hammer.

3. Turn the block over and cut the opposite side using the same procedure as in Step 2. When the cut is made, the block should break into two pieces.　　　Completed ☐

4. Practice cutting several blocks to determine the amount of force needed to make a quality cut. Completed ☐
5. Clean up the pieces and return all tools and materials to their proper places. Completed ☐

Instructor's Initials: _____

Date: _____

Job 15 Review

After completing this job successfully, answer the following questions.

1. What are the names of the tools used to cut concrete blocks by hand?
   _____
   _____

2. In what position should you hold the blocking chisel when cutting a concrete block?
   _____
   _____

3. What piece of safety equipment is absolutely essential when cutting concrete block with a blocking chisel?
   _____
   _____

4. Why is it necessary to score both sides of the concrete block?
   _____
   _____
   _____

5. Can a concrete block be cut down through the web with the blocking chisel?
   _____
   _____

Score: _____

Name _____ Date _____ Class _____

JOB 16: Using a Mason's Line

OBJECTIVE: After completing this job, you will be able to use a mason's line to establish the proper course height for a masonry wall.

TEXTBOOK REFERENCE: Study the appropriate sections in Chapter 12, *Laying Brick* and Chapter 13, *Laying Block,* before starting this job.

Equipment

To complete this job, you will need the following tools and materials:

- A length of mason's line
- Mason's rule
- Plumb rule (level)
- Two leads or stacks of bricks or blocks to serve as the lead (corners)
- Two line blocks (for bricks)

Recommended Procedure

1. Select a partner to work with you on this job. Using a pair of leads (corners) of brick masonry, stretch out a length of mason's line long enough to reach from the outside edge of one lead to the outside of the other. Completed ☐

2. Ask your instructor to show you how to attach the line to one of the line blocks. Have your partner hold the line block against the corner at the top edge of a course of masonry. Completed ☐

3. Go to the other lead and place a line block at the same height. Attach the mason's line to the line block. Pull the line taut until all sag has been removed. Completed ☐

4. Using the plumb rule (level), check the line to ensure it is level. Completed ☐

5. Using the mason's rule, measure the distance from the bed course of masonry to the location of the line block at each end. Compare the dimensions. Completed ☐

6. Repeat the procedure, but reverse roles with your partner. Try the process two or three times until you can perform the task efficiently. Completed ☐

7. Return the materials to their proper places. Completed ☐

TRADE TIP
This job can be repeated using adjustable line stretchers on concrete block leads.

Instructor's Initials: _____

Date: _____

Copyright Goodheart-Willcox Co., Inc.
May not be reproduced or posted to a publicly accessible website.

Job 16 Review

After completing this job successfully, answer the following questions.

1. When should a mason's line be used instead of a plumb rule only?

   _____
   _____
   _____

2. How could you tell when the mason's line was tight enough?

   _____
   _____
   _____

3. What devices support each end of a mason's line when a brick wall is being laid? What devices support the ends of a mason's line when concrete blocks are being laid?

   _____
   _____
   _____

4. How far away from the masonry units should the mason's line be positioned?

   _____
   _____

5. If a wall is very long, how is the mason's line supported in the middle to eliminate the sag?

   _____
   _____

   Score: _____

Name _____ Date _____ Class _____

JOB 17: Erecting Batter Boards

OBJECTIVE: After completing this job, you will be able to erect batter boards to preserve a foundation line using the proper technique.

TEXTBOOK REFERENCE: Study the appropriate section in Chapter 12, *Laying Brick* and Chapter 13, *Laying Block,* before starting this job.

Equipment

To complete this job, you will need the following tools and materials:

- Brick hammer
- A dozen 16 penny nails
- Four 2″ × 4″ × 2′ boards, sharpened on one end
- Two 1″ × 4″ × 3′ boards
- Two 1″ × 1″ × 12″ stakes
- Length of mason's line
- Plumb rule (level) or plumb bob
- 25′ flexible tape

Recommended Procedure

1. Batter boards are used to preserve the building line during excavation and construction. This job will establish a single building line. Drive the two 1″ × 1″ × 12″ stakes in the ground 12′ apart along an imaginary building line. See **Figure 17-1**. Completed ☐

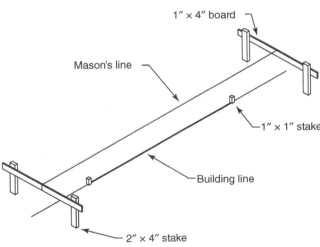

Figure 17-1. Batter boards locating a building line.

2. Move out 3′ beyond one stake, but in line with the building line. Drive two 2″ × 4″ × 2′ stakes into the ground several inches deep, with the widest side facing the building line stake. These stakes should be about 2′ apart. Completed ☐

3. Carefully nail a 1″ × 4″ × 3′ board across the 2″ × 4″ stakes about 12″ to 18″ above the ground. This board should be relatively level. Completed ☐

4. Move to the other end of the building line and repeat the operation.   Completed ☐

5. Stretch the line across the horizontal board at either end so it passes directly over the building line stakes. Tie it off with a slip knot. Use the level or plumb bob to be sure the line is over the stake.   Completed ☐

6. Mark the position of the line on the horizontal board at each end so it can be removed and then positioned at the same location again later.   Completed ☐

7. Remove the line, nails, and stakes and return all tools and materials to their assigned places.   Completed ☐

Instructor's Initials: _____

Date: _____

Job 17 Review

After completing this job successfully, answer the following questions.

1. What are batter boards used for?

   _____
   _____
   _____

2. What materials are generally used for the construction of batter boards?

   _____
   _____

3. Why is it necessary to mark the location of the line on the batter boards?

   _____
   _____
   _____

4. Is it necessary for the cross boards to be exactly level?

   _____
   _____

5. Why are the batter boards positioned well away from the wall they locate?

   _____
   _____
   _____

6. Why is a slip knot or bow used to tie off the line on a batter board?

   _____
   _____
   _____

Score: _____

Name _____ Date _____ Class _____

JOB 18: Laying Six Bricks on a Board

OBJECTIVE: After completing this job, you will be able to lay a course of bricks without a line using the proper tools and techniques.

TEXTBOOK REFERENCE: Study the appropriate section in Chapter 12, *Laying Brick*, before starting this job.

Equipment

To complete this job, you will need the following tools and materials:

- Mason's trowel
- Mortar board
- Mortar (provided by instructor)
- 6 bricks
- 6' board 2" × 4" (straight)
- 3 concrete blocks
- Five-gallon bucket with water
- Plumb rule (level)
- Mason's rule
- Brick hammer
- V-jointer or convex jointer
- Brush

Recommended Procedure

1. Assemble the required tools and materials and set up a workspace by placing the 2" × 4" board across two of the concrete blocks. This will raise the workspace above the floor and make your work easier. The board should be close to level. See **Figure 18-1**.   Completed ☐

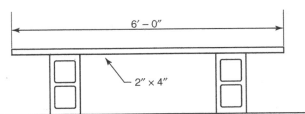

Goodheart-Willcox Publisher

Figure 18-1. The basic setup.

2. Stand your level in a cell hole in the remaining concrete block. Wet the mortar board and fill it with mortar. Place it on the five-gallon bucket.   Completed ☐

3. Lay the six bricks on the board with a 3/8" space between them. Measure each one with your rule. Mark the joint locations on the side of the board.   Completed ☐

4. Spread mortar on the board to form the bed joint for six bricks. Furrow the mortar bed and cut off excess mortar along the edges of the board.   Completed ☐

5. Place the first brick on the mortar bed at the end on your left as you face the board. Press the brick into the mortar so that a bed joint 3/8" thick is formed. This equals No. 6 on the modular rule. Level the brick and be sure it is parallel to the board.   Completed ☐

6. Form a head joint on the second brick and put it in place next to the first brick. The head joint should also be 3/8″ wide. Remove excess mortar. Completed ☐

7. Lay the remaining bricks along the board and remove all excess mortar. Check the course to be sure it is level, straight, and plumb. See **Figure 18-2**. Completed ☐

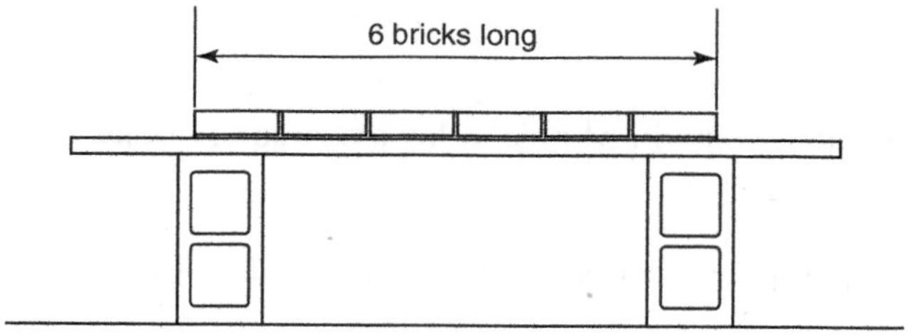

Goodheart-Willcox Publisher

Figure 18-2. The completed job.

8. When the mortar is thumbprint hard, strike the joints with the V-jointer or convex jointer. Remove any tags with the trowel. Completed ☐

9. Brush the brickwork to remove any excess mortar or dirt. Completed ☐

10. Clean up the work area and tools and return the tools and materials to their assigned places. Completed ☐

Instructor's Initials: _____

Date: _____

Job 18 Review

After completing this job successfully, answer the following questions.

1. What was the function of the board in this job?

   _____
   _____
   _____

2. Why was it necessary for the 2″ × 4″ board to be fairly close to level for this job?

   _____
   _____
   _____

3. Why was it recommended that you wet the mortar board before putting mortar on it?

   _____
   _____
   _____

4. What is the mortar joint thickness when a standard brick is used and laid to No. 6 on the modular rule?

   _____
   _____

Name _____

5. What tool was used to remove the tags or fins after the bricks were jointed?

_____

_____

6. What operation was performed immediately after a brick was laid?

_____

_____

Score: _____

Notes

Name _____ Date _____ Class _____

JOB 19: Laying a Four Course, Single Wythe, Running Bond Lead

OBJECTIVE: After completing this job, you will be able to lay a simple four course, single wythe, running bond lead (corner) using the proper technique.

TEXTBOOK REFERENCE: Study the appropriate section in Chapter 12, *Laying Brick,* before starting this job.

Equipment

To complete this job, you will need the following tools and materials:

- Mason's trowel
- Mortar board
- Mortar (provided by instructor)
- 14 bricks
- Five-gallon bucket with water
- Plumb rule
- Modular rule
- Chalk line
- Brick hammer
- Convex jointer and sled runner
- Brush

Recommended Procedure

1. Snap a chalk line where the face of the masonry wall is to be located along adjacent sides. Mark the actual head joints along the chalk line for proper placement of the masonry units. See **Figure 19-1**.

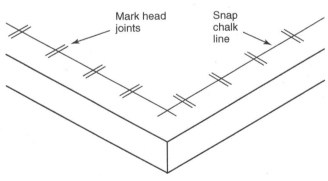

Goodheart-Willcox Publisher

Figure 19-1. Step 1.

2. Lay down a mortar bed and start the first course by laying the first unit from the corner. Level and plumb this brick. Remove excess mortar. Note that the placement marks for the head joints are visible even when the unit is in place. See **Figure 19-2**.

Completed

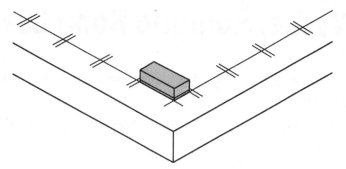

Figure 19-2. Step 2.

3. Complete the first course along both legs of the lead. Check to be sure that the bricks are properly spaced, straight, level, and plumb. Each leg should be about the same length. Leads are generally about seven or nine courses high, but a shorter lead may be used for practice, as in this case. See **Figure 19-3**.

Completed

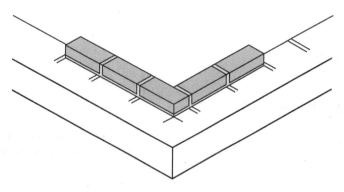

Goodheart-Willcox Publisher

Figure 19-3. Step 3.

4. Lay up the second and third courses, alternating the pattern of brick placement. Lay up the third course similar to the first course. See **Figure 19-4**.

Completed

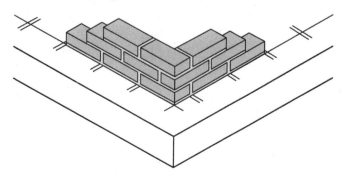

Goodheart-Willcox Publisher

Figure 19-4. Step 4.

Name _____

5. Lay up the fourth course to complete the lead. Two bricks on the highest course of the lead will provide enough resistance to maintain a tight line. See **Figure 19-5**. Completed ☐

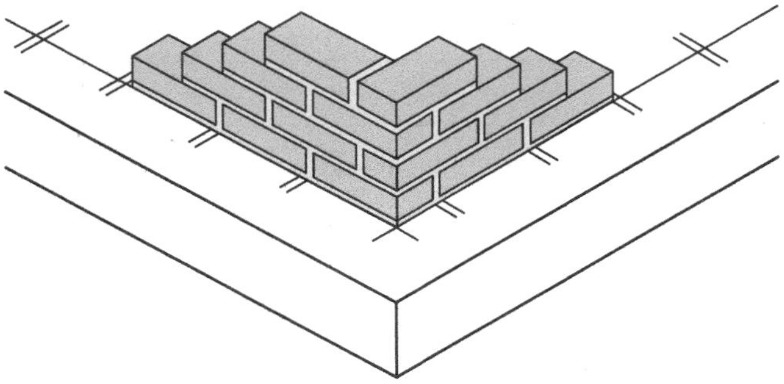

Goodheart-Willcox Publisher

Figure 19-5. Step 5.

6. Strike the joints when they are thumbprint hard. Remove any tags with the trowel and brush the wall to remove any mortar or dirt. Completed ☐

7. Clean up the work area and return all tools and materials to their assigned places. Completed ☐

Instructor's Initials: _____

Date: _____

Job 19 Review

After completing this job successfully, answer the following questions.

1. What is a *wythe*?

   _____
   _____
   _____

2. What mason's tool is generally used to straightedge a course of brick?

   _____
   _____

3. Where is the first brick laid when building a lead?

   _____
   _____

4. In a running bond, what courses are identical?

   _____
   _____

5. Approximately how many courses high are most brick leads laid?

   _____
   _____

6. Why are leads built?

   _____

   _____

   _____

7. If a wall is 10′ long and you do not want to build leads, what other approach could you use to support the mason's line?

   _____

   _____

   _____

   Score: _____

Name _____ Date _____ Class _____

JOB 20: Laying a 4″ Running Bond Wall with Leads

OBJECTIVE: After completing this job, you will be able to lay up a 4″ running bond wall with leads using the proper technique.

TEXTBOOK REFERENCE: Study the appropriate section in Chapter 12, *Laying Brick*, before starting this job. Note: This job follows the pictorial sequence in your text. Refer to the photos when needed.

Equipment

To complete this job, you will need the following tools and materials:

- Mason's trowel
- Mortar board
- Mortar (provided by instructor)
- Supply of bricks
- Five-gallon bucket with water
- Plumb rule
- Mason's rule
- Chalk line
- Brick hammer
- Jointer

Recommended Procedure

1. Arrange your workspace for efficiency. The mortar board should be located in the center of the workspace, about 24″ from the wall. Brick should be stacked on both sides of the board. Completed ☐

2. Establish a wall line 6′ long using a chalk line. Check each corner to be sure it is square. You can do this by measuring 6′ along one side and 8′ along the intersecting side. The diagonal distance between the points should be 10′. This procedure is known as the *3-4-5 method* (3-4-5 doubled is 6-8-10). Completed ☐

3. Lay out a dry course of bricks from corner to corner with uniform head joints. Mark the joints along the chalk line and then move the bricks aside. Completed ☐

4. Spread the mortar on one side of the corner and furrow it. Lay the corner brick exactly on the point where the corner is located. It must be set level and square with the wall line. Completed ☐

5. Lay the remaining four or five bricks of the lead corner. This is called "tailing out" the lead of the corner. After the bricks have been laid, level them with the plumb rule. Plumb the corner brick and then the tail end. Leveling is done on the outside and top edge of the brick. Completed ☐

6. Line up the bricks between the two plumb points. Here the level is used as a straightedge and the bubbles are disregarded. Always follow this sequence when building a corner—level the unit, then plumb, and then line up. Completed ☐

7. After one side of the corner has been laid and trued, start the other side of the corner. Spread the mortar and furrow it. Lay three or four bricks. Level, plumb, and line them up. Do not tap the level with the trowel or hammer—use your hand. This time the corner brick does not require plumbing because it has already been plumbed. Completed ☐

8. Lay the second course in the same sequence as the first course. Check for proper height just after the course has been laid, but before the bricks have been leveled. If the bricks are too high, tap them down as they are leveled. If they are too low, remove them and add more mortar before leveling. Completed ☐

9. Repeat the sequence until the corner is built to a height of seven courses.   Completed ☐

10. To eliminate any wind, belly, or cave-in in the wall, straightedge the rack of the lead by laying the level across the corner of each brick in the rack. The courses should line up.   Completed ☐

11. Lay the second corner of the wall following the same procedure that was used for the first corner.   Completed ☐

12. Stretch the mason's line between the corners at the top of the first course. Use corner blocks and pull the line taut enough to remove any sag. Use the same amount of tension each time.   Completed ☐

13. Begin laying the wall from the lead toward the center, one course at a time. Each brick on the second course should be centered over the cross joint of the first course. Be sure the head joints are uniform so the last brick will fit properly. The line should be worked from both leads toward the center. Level, plumb, and line up each course as you lay it.   Completed ☐

14. Tool or strike the joints when the mortar is thumbprint hard. Use the sled jointer (long jointer) for the bed joints and then the short jointer for the vertical joints. Remove any tags with the trowel. You may want to use the jointers again to improve appearance.   Completed ☐

15. Brush the wall when the mortar is stiff enough. Brushing reduces the amount of cleaning required later.   Completed ☐

16. Lay the remaining courses in the same manner until the wall is completed to the required height of seven courses.   Completed ☐

17. Examine your work and note where it could be improved.   Completed ☐

18. Clean up the area and remove all mortar from your tools. Return materials and tools to their assigned places.   Completed ☐

Instructor's Initials: _____

Date: _____

Job 20 Review

After completing this job successfully, answer the following questions.

1. How is a corner squared using the 3-4-5 method?

   _____

   _____

   _____

2. Why is a dry course of brick laid out before any mortar is used?

   _____

   _____

   _____

3. What is the proper order for the following operations when laying a course of bricks (line up the bricks, level the bricks, plumb the bricks)?

   _____

   _____

4. If a brick has too thick a mortar bed, how should you tap it down to the proper level?

   _____

   _____

Name _____

_____

5. What is the recommended procedure when a brick is too low (mortar bed is too thin)?

_____
_____
_____

6. How is the rack straightedged?

_____
_____
_____

Score: _____

Notes

Name _____ Date _____ Class _____

JOB 21: Laying an 8" Common Bond, Double Wythe Brick Wall with Leads

OBJECTIVE: After completing this job, you will be able to lay up an 8" common (American) bond, double wythe brick wall with leads using the proper technique.

TEXTBOOK REFERENCE: Study the appropriate section in Chapter 12, *Laying Brick*, before starting this job.

Equipment

To complete this job, you will need the following tools and materials:
- Mason's tools
- Mason's line
- Mortar
- Supply of bricks

Recommended Procedure

1. Lay out the wall location as you did with the 4" wall in Job 20. Locate a second line inside the first wall line equal to the length of the bricks being used (about 8"). Your instructor will specify the length of the wall and lead tails.   Completed ☐

2. Lay out the bond to eliminate cutting, if possible. This is an important step. Mark the head joints to ensure proper and uniform placement.   Completed ☐

3. Lay the corner bricks of the lead as shown in **Figure 21-1**. Level, plumb, and square these bricks.   Completed ☐

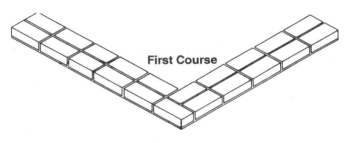

Goodheart-Willcox Publisher

Figure 21-1. First course.

4. Complete the first course following the pattern illustrated in **Figure 21-1**. Four or five stretchers should be sufficient. Check with your instructor.   Completed ☐

5. Study **Figure 21-2** before laying the second course of the lead. Notice that you will need 3 three-quarter closures, 1 bat (half brick), and 1 closure. In this job, the header course is the second course, but it could begin wherever specified between five to seven courses of stretchers.

Completed ☐

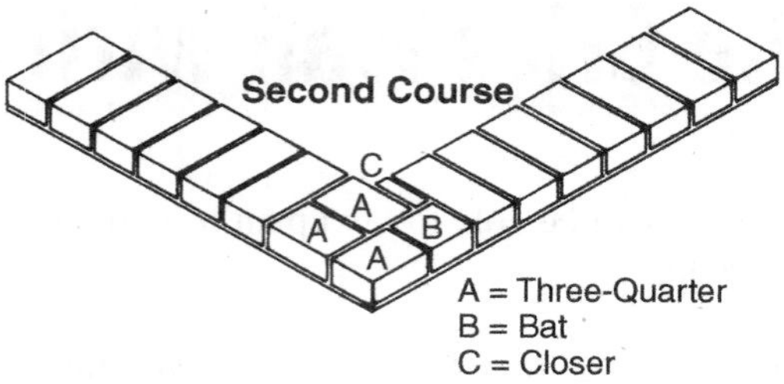

A = Three-Quarter
B = Bat
C = Closer

Goodheart-Willcox Publisher

Figure 21-2. Second course.

6. Lay the third through seventh courses of stretchers. Level, plumb, and line up each course as it is laid. See **Figure 21-3**.

Completed ☐

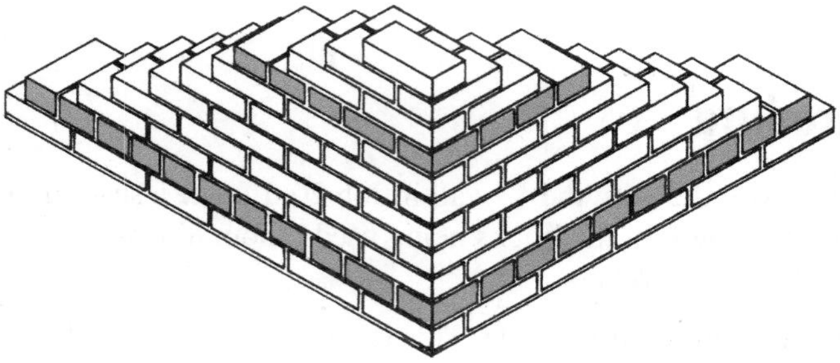

Goodheart-Willcox Publisher

Figure 21-3. Completed corner lead.

7. Lay the eighth course of headers using the same pattern that was used in the second course. Notice how this configuration breaks the bond of the previous course.

Completed ☐

8. Finish laying the first corner as shown in **Figure 21-3**. Strike all the joints as the mortar hardens to thumbprint hard. Remove the tags and brush the bricks. This is a good place to stop for the day.

Completed ☐

9. Lay the second corner using the same procedure that you followed in laying up the first corner. Refer to **Figure 21-1**, **Figure 21-2**, and **Figure 21-3**.

Completed ☐

Name _____

10. Construct the first stretcher course of the wall by laying the outside first and then the inside. These should be laid from the leads toward the center. Level, plumb, and line up each course before beginning the next course. Refer to **Figure 21-4**. Completed ☐

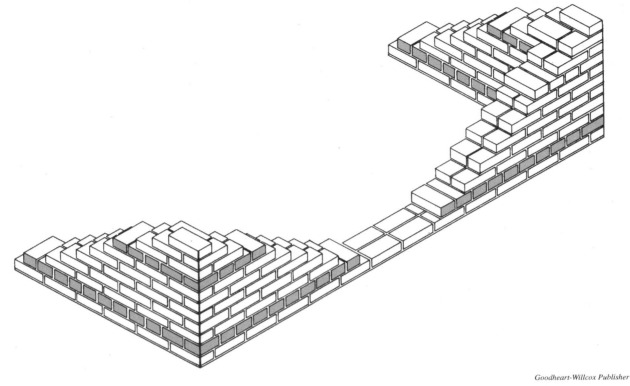

Goodheart-Willcox Publisher

Figure 21-4. Corner leads with first stretcher course.

11. Use a line to establish the height of the header course. Lay the header (second) course between the leads. Watch the width of your mortar joints to be sure the proper spacing is maintained. Level, plumb, and line up the header course. Completed ☐

12. Lay the outside tier up to the next header course, keeping it straight, level, and plumb. This is called "header high." Use the mason's line. Completed ☐

13. Lay the inside tier up to the same height. Be careful to keep it level with the outside tier as it is laid up. This is important, as you must have a level surface for the header course. Completed ☐

14. Continue as before until the desired height is reached. Strike the joints and brush the wall at the proper time. Completed ☐

15. Inspect your work, clean up the area, and return the unused materials and tools to their proper places. Completed ☐

Instructor's Initials: _____

Date: _____

Job 21 Review

After completing this job successfully, answer the following questions.

1. How can a common bond wall be recognized?

   _____
   _____
   _____

2. In addition to headers and stretchers, what other bricks were used to build this wall?

   _____
   _____

3. Once the first lead is built, what is the next operation in building an 8″ common bond, double wythe wall with leads?

   _____
   _____
   _____

4. How should the bricks be laid between the leads?

   _____
   _____
   _____

5. Should you wait until the wall is built before striking any joints?

   _____
   _____

6. What tool should be used to strike the bed joints on a brick wall?

   _____

   Score: _____

Name _____ Date _____ Class _____

JOB 22: Laying an 8″ Running Bond, Two-Wythe Intersecting Brick Wall

OBJECTIVE: After completing this job, you will be able to lay an 8″ running bond, two-wythe intersecting brick wall using the proper technique.

TEXTBOOK REFERENCE: Study the appropriate section in Chapter 12, *Laying Brick,* before starting this job.

Equipment
To complete this job, you will need the following tools and materials:
- Mason's tools
- Mortar
- Supply of bricks
- Chalk line
- Metal Z ties

Recommended Procedure
Intersecting brick walls are often necessary in the construction of brick structures. They are laid out so the courses are tied together to form an integrated unit. The individual units are placed so they interlock the wall segments together. Reinforcement ties (Z ties) are generally used.

1. Study the first course configuration in **Figure 22-1**. Locate the face of the walls on the floor and snap a chalk line to preserve the location.   Completed ☐

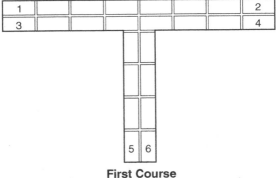

Goodheart-Willcox Publisher

Figure 22-1. First course.

2. Lay out the first course using no mortar (dry bond) to maintain proper spacing and identify any difficulties or problems. See **Figure 22-1**.   Completed ☐

3. Spread the mortar bed and lay the face wythe beginning with brick #1 through brick #2 shown in **Figure 22-1**. Level, align, and plumb the course of bricks.   Completed ☐

4. Lay the backing wythe beginning with brick #3 and continuing to brick #4. Do not forget to butter the back side of each brick where the backing wythe sits against the face course. This joint (collar joint) should be completely filled. Level, align, and plumb the course and check the wall thickness.   Completed ☐

Copyright Goodheart-Willcox Co., Inc.
May not be reproduced or posted to a publicly accessible website.

5. Lay the intersecting wall brick beginning with brick #5 and then brick #6, running each course to the outside wall. Level, align, and plumb each course and check the wall thickness. This will complete the first course plan. Completed ☐

6. Lay the second course following the layout shown in **Figure 22-2**. Level, align, and plumb the course. Notice how the intersecting wall is interlocked with the outside wall. Completed ☐

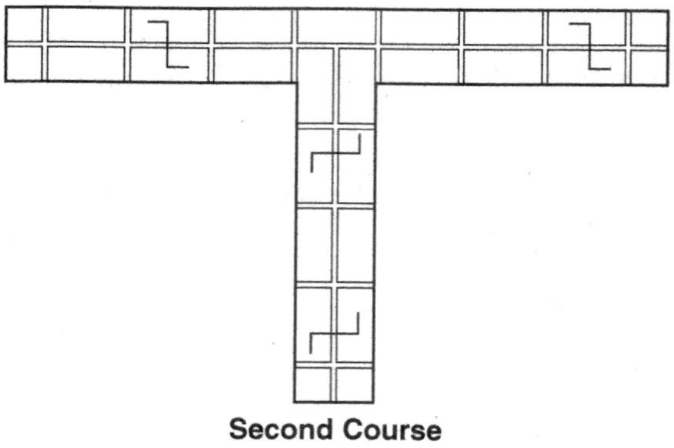

Second Course

Goodheart-Willcox Publisher

Figure 22-2. Second course.

7. Locate the metal Z ties as shown in **Figure 22-2**. Embed the metal ties in a mortar bed and lay the third course identical to the first. Metal ties will be added again on the eighth course in the wall. Completed ☐

8. Continue laying courses alternating the pattern used in the first and second courses until the wall has reached eight courses high. Refer to **Figure 22-3**. Completed ☐

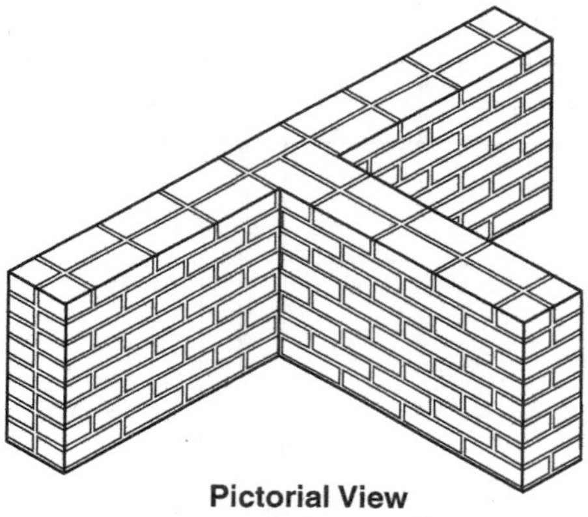

Pictorial View

Goodheart-Willcox Publisher

Figure 22-3. Pictorial view.

9. Level and plumb the completed wall and clean off any mortar splatter. Completed ☐

10. Tool the joints with a concave jointer when the mortar is thumbprint hard. Tool the head joints first, then the bed joints. Use the sled jointer for the bed joints. Completed ☐

11. Remove any mortar tailings (residue) that remain after finishing the joints. Clean the wall by brushing with a bricklayer's brush. Completed ☐

12. Clean up the area and return any unused materials and tools to their proper places. Completed ☐

Instructor's Initials: _____

Date: _____

174   Modern Masonry Lab Workbook

Name _____

Job 22 Review

After completing this job successfully, answer the following questions.

1. What kind of metal ties are generally used to tie the two wythes together in an 8″ running bond, two-wythe wall?
   _____
   _____

2. In addition to the metal ties, how are the wythes bonded together?
   _____
   _____
   _____

3. How is the intersecting wall bonded to the outside wall?
   _____
   _____
   _____

4. Which joints, head joints or bed joints, are tooled first?
   _____

5. What is the operation that follows removing the fins or tailings?
   _____
   _____

                                    Score: _____

Notes

Name _____ Date _____ Class _____

JOB 23: Building a Corner in Flemish Bond with Quarter Closures

OBJECTIVE: After completing this job, you will be able to build a corner in Flemish bond with quarter closures using the proper technique.

TEXTBOOK REFERENCE: Study the appropriate section in Chapter 12, *Laying Brick*, before starting this job.

Equipment

To complete this job, you will need the following tools and materials:
- Mason's tools
- Mortar
- Supply of bricks

Recommended Procedure

1. The popular Flemish bond produces an artistic and pleasing wall. It is more costly than the common bond and requires greater care, but is worth the effort. See **Figure 23-1** for the pattern. Completed ☐

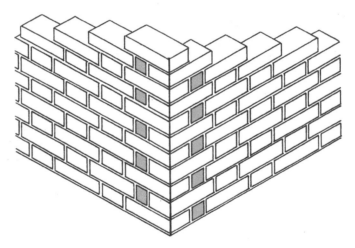

Goodheart-Willcox Publisher

Figure 23-1. Flemish bond corner with a quarter closure.

2. The bond consists of alternate headers and stretchers in each course. The headers are centered on the stretchers between each course. The bond is started at the corner with a stretcher and then a quarter closure along the opposite tail. Lay the first course of brick. Completed ☐

3. Lay the remaining courses following the pattern shown in **Figure 23-1**. Your instructor will specify the height of the corner. Completed ☐

4. Tool the joints and brush the bricks. Completed ☐

5. Inspect your work, clean up the area, and return unused materials and tools to their assigned places. Completed ☐

Instructor's Initials: _____

Date: _____

Job 23 Review

After completing this job successfully, answer the following questions.

1. How does the Flemish bond with quarter closures get its name?

   _____
   _____
   _____

2. Why is the Flemish bond with quarter closures more costly?

   _____
   _____
   _____

3. Which courses have headers?

   _____
   _____

4. Why is the Flemish bond popular?

   _____
   _____

5. When should the wall be brushed?

   _____
   _____

   Score: _____

Name _____ Date _____ Class _____

JOB 24: Building a Corner in Flemish Bond with Three-Quarter Closures

OBJECTIVE: After completing this job, you will be able to build a corner in Flemish bond with three-quarter closures using the proper technique.

TEXTBOOK REFERENCE: Study the appropriate section in Chapter 12, *Laying Brick*, before starting this job.

Equipment

To complete this job, you will need the following tools and materials:
- Mason's tools
- Mortar
- Supply of bricks

Recommended Procedure

Laying up a corner or wall in Flemish bond using three-quarter closures is identical to building a wall using quarter closures (Job 23), except that three-quarter closures are used instead of quarter closures.

1. The Flemish bond using three-quarter closures is a popular bond. Refer to **Figure 24-1** to see the pattern. Study the drawing before beginning work. Completed ☐

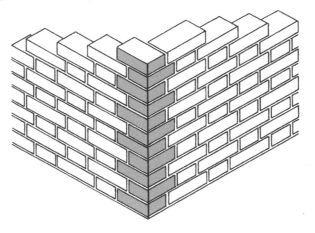

Goodheart-Willcox Publisher

Figure 24-1. Flemish bond with three-quarter closures at the corner.

2. The bond consists of alternate headers and stretchers in each course. The headers are centered on the stretchers between each course. The bond is started at the corner with a three-quarter closure and then a stretcher along the opposite tail. Lay the first course of bricks. Completed ☐

3. Lay the remaining courses following the pattern shown in **Figure 24-1**. Your instructor will specify the height of the corner. Completed ☐

4. Tool the joints and brush the bricks. Completed ☐

5. Inspect your work, clean up the area, and return unused materials and tools to their assigned places. Completed ☐

Instructor's Initials: _____

Date: _____

Job 24 Review

After completing this job successfully, answer the following questions.

1. How is the Flemish bond with three-quarter closures different from the Flemish bond with quarter closures?

   _____
   _____
   _____

2. Describe the brick pattern in a corner laid in Flemish bond with three-quarter closures.

   _____
   _____
   _____

3. What are the disadvantages of using the Flemish bond with three-quarter closures?

   _____
   _____
   _____

4. What are the advantages of a Flemish bond wall with three-quarter closures?

   _____
   _____
   _____

5. What are the pieces of brick left over from cutting the three-quarter closures called?

   _____

   Score: _____

Name _____ Date _____ Class _____

JOB 25: Building a Corner in English Bond with Quarter Closures

OBJECTIVE: After completing this job, you will be able to build a corner in English bond with quarter closures using the proper technique.

TEXTBOOK REFERENCE: Study the appropriate section in Chapter 12, *Laying Brick,* before starting this job.

Equipment

To complete this job, you will need the following tools and materials:

- Mason's tools
- Mortar
- Supply of bricks

Recommended Procedure

1. The English bond has alternate courses of stretchers and headers. The headers center on the stretchers and on the joints between the stretchers. The stretchers all line up vertically, one over the other. See **Figure 25-1** for the pattern.   Completed ☐

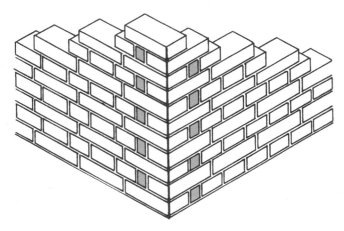

Goodheart-Willcox Publisher

Figure 25-1. English bond with a quarter closure.

2. The bond is started at the corner with a stretcher along one tail and a quarter closure along the other tail. Notice that stretchers are continued along one side while headers are used along the other side. Lay the first course.   Completed ☐

3. Reverse the pattern for the second course. Lay the second course.   Completed ☐

4. Continue laying up the wall alternating courses as shown in **Figure 25-1** until the desired height is reached.   Completed ☐

5. Tool the joints and brush the bricks.   Completed ☐

6. Inspect your work, clean up the area, and return unused materials and tools to their assigned places.   Completed ☐

Instructor's Initials: _____

Date: _____

Copyright Goodheart-Willcox Co., Inc.
May not be reproduced or posted to a publicly accessible website.

Job 25 Review

After completing this job successfully, answer the following questions.

1. What is the most dominant pattern in an English bond with quarter closures?
   _____
   _____

2. How are the headers positioned with respect to the stretchers in an English bond with quarter closures?
   _____
   _____

3. What is the pattern of the stretchers in an English bond with quarter closures?
   _____
   _____

4. Is this bond practical for a single-wythe wall? Why?
   _____
   _____
   _____

5. Is it possible to build an English bond wall without any quarter closures?
   _____
   _____
   _____

Score: _____

Name _____ Date _____ Class _____

JOB 26 — Building a Corner in English Bond with Three-Quarter Closures

OBJECTIVE: After completing this job, you will be able to build a corner in English bond with three-quarter closures using the proper technique.

TEXTBOOK REFERENCE: Study the appropriate section in Chapter 12, *Laying Brick,* before starting this job.

Equipment

To complete this job, you will need the following tools and materials:

- Mason's tools
- Mortar
- Supply of bricks

Recommended Procedure

1. English bond has alternate courses of stretchers and headers. The headers center on the stretchers and on the joints between the stretchers. The stretchers all line up vertically, one over the other. See **Figure 26-1** for the pattern.   Completed ☐

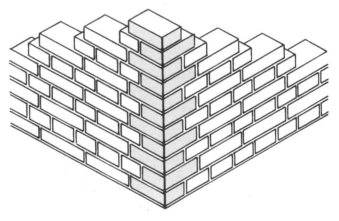

Goodheart-Willcox Publisher

Figure 26-1. English bond with three-quarter closures.

2. The bond is started at the corner with a three-quarter closure along one tail and a header along the other tail. Notice that stretchers are continued along one side, while headers are used along the other side. Lay the first course.   Completed ☐

3. Reverse the pattern for the second course. Lay the second course.   Completed ☐

4. Continue laying up the wall alternating courses as shown in **Figure 26-1** until the desired height is reached.   Completed ☐

5. Tool the joints and brush the bricks.   Completed ☐

6. Inspect your work, clean up the area, and return unused materials and tools to their assigned places.   Completed ☐

Instructor's Initials: _____

Date: _____

Job 26 Review

After completing this job successfully, answer the following questions.

1. How does a corner in English bond with three-quarter closures differ from one with quarter closures?

   _____
   _____
   _____
   _____

2. Is English bond practical for a single-wythe wall? Why?

   _____
   _____
   _____
   _____

3. In a four-course corner in English bond with three-quarter closures, how many of the courses have headers?

   _____

4. How does the appearance of an English bond wall with three-quarter closures compare with an English bond wall with quarter closures (not the corners)?

   _____
   _____
   _____
   _____

5. Would English bond be more practical if the wall were an 8″ solid brick wall? Why?

   _____
   _____
   _____
   _____

                                        Score: _____

Name _____ Date _____ Class _____

JOB 27: Constructing a 10" Brick Masonry Cavity Wall with Metal Ties and Weep Holes

OBJECTIVE: After completing this job, you will be able to construct a 10" brick masonry cavity wall with metal ties and weep holes using the proper technique.

TEXTBOOK REFERENCE: Study the appropriate section in Chapter 12, *Laying Brick,* before starting this job.

Equipment
To complete this job, you will need the following tools and materials:
- Mason's tools
- Chalk line
- Mortar
- Supply of bricks
- Metal Z ties
- Four 6" oiled rods or wick
- One drop stick 1 3/4" × 3/4" × 48" long
- Eye protection for cutting bricks

Recommended Procedure
No changes are required in basic bricklaying techniques in the construction of a cavity wall. However, no bridge of solid material capable of carrying water across the minimum 2" cavity space is permitted. The construction of two separate wythes, with a clean cavity, is the objective.

1. Study **Figure 27-1** before starting work on the cavity wall.   Completed ☐

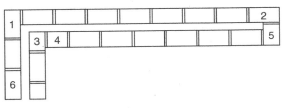

Goodheart-Willcox Publisher

Figure 27-1. First course.

2. Snap chalk lines to establish the wall location. Determine the length of both sections of the wall by counting the bricks in **Figure 27-1**.   Completed ☐

3. Set up the work area with two mortar boards, one on either side of the wall. Stack bricks on either side of both mortar boards.   Completed ☐

4. Lay out the first course dry as shown in the first course plan and mark the position of each brick.   Completed ☐

5. Lay the inside bed course beginning with brick #1 and continuing to brick #2 in a full mortar bed. Work from both ends to the middle. Level and plumb these bricks. The bricks should be laid to #6 on the modular rule. Lay brick #6 and fill in the course between #1 and #6.   Completed ☐

6. Lay brick #5 and the one adjacent to it. Lay bricks #3 and #4 to the line. Be sure all corners are square. Fill in the bricks to complete the first course. Check your work to be sure it is level, plumb, and straight.   Completed ☐

7. If you did not use empty head joints for the four weep holes shown in the elevation plan, **Figure 27-2**, open those joints or use any one of the methods of forming a weep hole—oiled rod, rope wick, or metal or plastic tubing. Be sure the two tiers of masonry are the same height and level.

Completed ☐

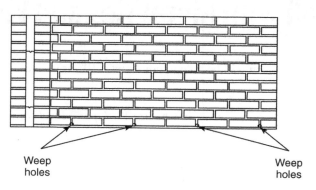

Figure 27-2. Elevation.

> **WARNING!**
> Be sure to wear safety glasses during the following procedure.

8. Cut 3 three-quarter brick for the second course and lay the mortar bed for the second course. Bevel the mortar bed to avoid dropping mortar into the cavity. Lay the second course of brick as shown in **Figure 27-3**. Lay three-quarter brick #7 at the right end of the wall over bed bricks #2 and #5. Note that the cross joints are centered on the bricks below. Keep the cavity clean. Lay the second course brick #8 at the other end of the wall and complete the wythe between them. Complete the second course.

Completed ☐

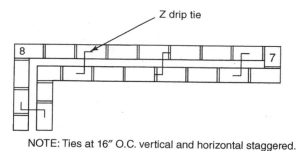

NOTE: Ties at 16" O.C. vertical and horizontal staggered.

Figure 27-3. Second course.

9. Install the Z ties in the position shown in **Figure 27-3**. The ties should be set in a mortar bed with the drip pointing down.

Completed ☐

10. Lay the third course, being careful not to drop any mortar into the cavity. Install the drop stick on the metal ties, moving it along to prevent mortar from entering the cavity.

Completed ☐

11. Lay the fourth through eighth courses, making sure they are level and straight. Remove the drop stick and insert the Z ties on top of the eighth course as shown in **Figure 27-3**.

Completed ☐

12. Install the drop stick on the metal ties and lay the next four courses to complete the wall. Be sure all courses are laid to No. 6 on the modular rule.

Completed ☐

13. Strike the mortar joints when the mortar is hard enough. Use the sled runner on the bed joints and the convex tool on the head joints. Brush the wall.

Completed ☐

Name _____

14. Check to see that the weep holes are clear. Completed ☐
15. Inspect your work, clean up the area, and return all unused materials and tools to their assigned places. Completed ☐

Instructor's Initials: _____

Date: _____

Job 27 Review

After completing this job successfully, answer the following questions.

1. Why must the cavity be kept clean of mortar droppings in a cavity wall?

   _____
   _____
   _____

2. At what point does an *air space* become a *cavity*?

   _____
   _____
   _____

3. Why are two mortar boards generally recommended for cavity wall construction?

   _____
   _____
   _____

4. What is the purpose of a weep hole?

   _____
   _____
   _____

5. Describe the basic types of weep holes.

   _____
   _____
   _____

6. What is the function of Z ties in a 10″ brick cavity wall?

   _____
   _____
   _____

Score: _____

Notes

Name _____ Date _____ Class _____

JOB 28: Constructing a Reinforced Single Wythe Brick Bearing Wall

OBJECTIVE: After completing this job, you will be able to construct a reinforced single wythe brick bearing wall and install steel reinforcing using the proper technique.

TEXTBOOK REFERENCE: Study the appropriate section in Chapter 12, *Laying Brick,* before starting this job.

Equipment

To complete this job, you will need the following tools and materials:

- Mason's tools
- Chalk line
- Mortar
- Supply of bricks
- Joint reinforcement
- 2′ pieces of No. 4 rebar

Recommended Procedure

In a single wythe brick bearing wall system, the brick masonry serves as both the structural system and exterior facing. Study **Figure 28-1**. The wall you are to build in this job is a 6″ reinforced wall made from 6″ × 4″ × 12″ face brick. The wall has vertical reinforcing of No. 4 rebar at 48″ O.C. and horizontal 3/16″ diameter reinforcing at 24″ O.C.

Goodheart-Willcox Publisher

Figure 28-1. A single wythe brick bearing wall.

1. Set up your work area and snap a line on the floor to maintain the location of the wall. The length of the wall should be about 48″ long. Ask your instructor if this length is satisfactory.   Completed ☐

2. Lay down a full mortar bed to receive the bed course. Make sure that you do not cover the chalk line with mortar. All joints must be full to resist moisture penetration.   Completed ☐

3. Lay the bed course in running bond, keeping the head joints even in thickness and completely filled. Level, plumb, and straightedge the course. Check spacing with your plumb rule.   Completed ☐

4. Lay the second course so that mortar joints are centered on the bricks in the first course. The cell holes must line up to receive reinforcing and mortar (or grout).   Completed ☐

5. Lay the third course, using the same spacing as the first course. Level, plumb, and straightedge the wall. Tool the mortar joints when the mortar is thumbprint hard. Completed ☐

6. Select a joint reinforcement that is compatible with the brick masonry units being used. Position the joint reinforcement on the top of the third course and embed it in mortar. Completed ☐

7. Lay the fourth course of bricks similar to the second course spacing. Make sure the reinforcing is completely embedded in mortar and remains in the proper position. Completed ☐

8. Lay the fifth course of bricks similar to the first and third courses. Insert the vertical reinforcing bar (maximum size No. 6) in a cell near the center of the wall. Completed ☐

9. Fill the cells containing the vertical reinforcing with mortar. The mortar should be fluid enough to fill the voids, but it should not separate into its constituent parts. Completed ☐

10. Strike the joints and brush the wall. Completed ☐

11. Inspect your work, clean up the area, and return unused materials and tools to their proper places. Completed ☐

Instructor's Initials: _____

Date: _____

Job 28 Review

After completing this job successfully, answer the following questions.

1. What is the difference between a wythe and a course of bricks?

   _____
   _____
   _____

2. What are the two functions served in a single wythe brick bearing wall system?

   _____
   _____
   _____

3. Is it always necessary to have both horizontal as well as vertical reinforcing in a single wythe brick bearing wall? Why?

   _____
   _____
   _____

4. Why was it necessary for the cell holes to line up in the wall you built?

   _____
   _____
   _____

5. What is the maximum recommended size of vertical reinforcing to be used in a reinforced single wythe brick bearing wall?

   _____

Score: _____

JOB 29: Corbeling a 12″ Brick Wall

OBJECTIVE: After completing this job, you will be able to corbel a brick wall using the proper technique.

TEXTBOOK REFERENCE: Study the appropriate section in Chapter 12, *Laying Brick*, before starting this job.

Equipment
To complete this job, you will need the following tools and materials:
- Mason's tools
- Chalk line
- Mortar
- Supply of bricks

Recommended Procedure

Masons corbel a wall to widen it by projecting out masonry units to form a ledge or shelf. Each brick course extends out farther than the one below it. As a general rule, a masonry unit should not extend out farther than one-third the width or one-half the height, whichever is less.

Corbels normally support a load and therefore must be carefully constructed. Headers are generally used to tie the corbel into the base. Building codes generally require the top course to be a full header course. All joints must be completely filled with mortar.

1. Study **Figure 29-1**. Snap a line on the foundation or base where the front edge of the wall is to be constructed. Completed ☐

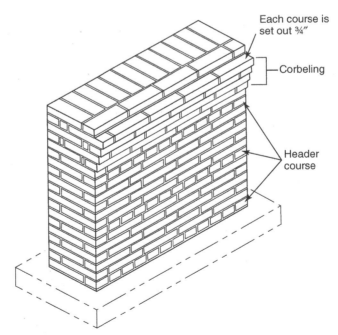

Figure 29-1. Corbeling a brick wall.

2. Lay out the first course as a dry course to check the spacing. Running bond will be used with a header course every seventh course.   Completed ☐

3. Lay the first course with headers on the front wythe and stretchers on the back. Use a full mortar bed. Level, plumb, and square the bricks. Check the spacing.   Completed ☐

4. Lay the second through sixth courses. Check the height of each course as it is laid and be sure bed joints are uniform and level.   Completed ☐

5. Lay the seventh course as a header course with full mortar joints.   Completed ☐

6. Continue to lay the wall until you reach the height where the corbel course is to begin. Check your progress often to be sure it is plumb and straight.   Completed ☐

7. Begin the corbel course by projecting headers out 3/4″ beyond the course below. Fill the extra wide head joint with mortar.   Completed ☐

8. Lay the second corbel course using stretchers along the front and headers along the back wythe. Fill in the space between with bats. Be sure the course is level and straight.   Completed ☐

9. Lay the third corbel course using stretchers on the front and back wythes and three-quarter headers between the wythes. Each corbel should project 3/4″ beyond the course below.   Completed ☐

10. Lay the next course in the same manner as the first course of the wall—headers on the front wythe and stretchers on the back wythe. Continue the wall to the desired height.   Completed ☐

11. Finish all joints when the mortar is hard enough to tool.   Completed ☐

12. Clean the wall with the trowel and brush.   Completed ☐

13. Inspect your work, clean up the area, and return tools and unused materials to their assigned places.   Completed ☐

Instructor's Initials: _____

Date: _____

Job 29 Review

After completing this job successfully, answer the following questions.

1. What is the function of a corbel in a masonry wall?

   _____

   _____

   _____

2. When a wall is corbelled, what is the maximum distance that a brick can be extended out beyond the unit below it?

   _____

3. What bonding technique is generally used to tie the corbel to the base?

   _____

4. What is important to remember about mortar joints in corbels?

   _____

   _____

   _____

Score: _____

Name _____ Date _____ Class _____

JOB 30: Building a Hollow Brick Masonry Pier

OBJECTIVE: After completing this job, you will be able to build a 16″ × 20″ hollow brick masonry pier using the proper technique.

TEXTBOOK REFERENCE: Study the appropriate section in Chapter 12, *Laying Brick,* before starting this job.

Equipment

To complete this job, you will need the following tools and materials:
- Mason's tools
- Chalk line
- Mortar
- Supply of bricks

Recommended Procedure

Piers are similar to columns, except they are shorter and generally do not support a load. Piers are commonly used as gateposts at corners or openings or at ends of a wall. Piers can be constructed using a one-wythe wall 4″ thick. The bond pattern is usually staggered so that the wall is tied together from a different side in each course in an interlocking fashion. Weep holes may be required. Study **Figure 30-1**.

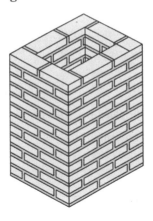

Figure 30-1. Pictorial view.

1. Study **Figure 30-2** to see the pattern used in the first course. Snap a line to preserve the location of the pier.   Completed ☐

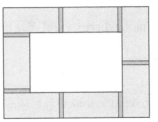

Figure 30-2. First course.

2. Lay the bed course on a generous mortar bed. Check the bricks to be sure they are level and straight and that the corners are square. You may use either 1/2″ or 3/8″ joints, depending on the type of bricks used. Refer to **Figure 30-2**. Completed ☐

3. Lay the second course as shown in the second course plan, **Figure 30-3**. Position bricks so head joints are offset to provide an interlocking connection. Level, plumb, and square the course. Completed ☐

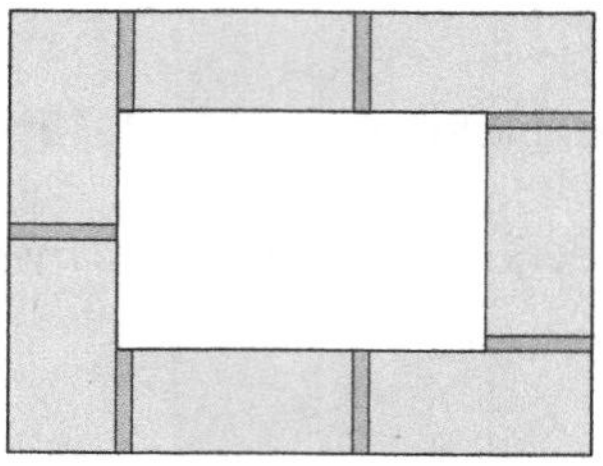

Goodheart-Willcox Publisher

Figure 30-3. Second course.

4. Lay successive courses, alternating the patterns used in the first and second courses until the pier has reached the desired height. Twelve courses should be adequate for practice. Avoid dropping any mortar inside the pier. Completed ☐

5. When the mortar is thumbprint hard, tool the joints and clean off any fins with the trowel. Clean the surface with a bricklayer's brush. Completed ☐

6. Inspect your work, clean up the area, and return any unused materials and tools to their assigned places. Completed ☐

Instructor's Initials: _____

Date: _____

Job 30 Review

After completing this job successfully, answer the following questions.

1. What is the difference between a pier and a column?

   _____
   _____
   _____

2. How did the pier that you constructed gain its strength since no reinforcing was used?

   _____
   _____
   _____

3. If weep holes had been required in your pier, where should they have been located?

   _____
   _____

4. How did you know it was time to tool (strike) the joints?

   _____
   _____

Name _____

5. What tool did you use to check the first course of masonry to be sure the pier was square?
   _____

6. How are fins generally removed from brick masonry?
   _____
   _____

                              Score: _____

Notes

Name _____ Date _____ Class _____

JOB 31 / Cleaning New, Dark Colored Brick Masonry

OBJECTIVE: After completing this job, you will be able to clean new, dark colored brick masonry with and without acid solutions using the proper technique.

TEXTBOOK REFERENCE: Study the appropriate section in Chapter 12, *Laying Brick,* before starting this job.

Equipment

To complete this job, you will need the following tools and materials:

- Eye and skin protection
- Long-handled stiff-fiber brush
- Mixing bucket (plastic)
- Wooden paddle or scraper
- Trisodium phosphate
- Household detergent
- Hydrochloric (muriatic) acid

Recommended Procedure

The appearance of a masonry structure may be ruined by improper cleaning. All cleaning should be applied to a sample test area of approximately 20 sq ft. It usually takes a minimum of one week to see the results of cleaning.

A hydrochloric acid solution is used extensively as a cleaning agent for new masonry, but steps 1 to 5 detail the procedure for cleaning without acid. When the masonry is thoroughly set and cured, begin the cleaning operation.

1. Remove large particles of mortar with wooden paddles or scrapers before wetting the wall. A chisel or wire brush might be necessary.   Completed ☐

2. Saturate the wall with clean water and flush away all loose mortar and dirt.   Completed ☐

3. Scrub the wall with a solution of 1/2 cup trisodium phosphate and 1/2 cup household detergent dissolved in onegallon of clean water. Use a stiff-fiber brush.   Completed ☐

4. Rinse off all cleaning solution and mortar particles using clean water under pressure.   Completed ☐

5. When acid cleaning becomes necessary on dark colored bricks, follow steps 1 and 2 above, and then use a clean, stain-free commercial grade hydrochloric (muriatic) acid. Mix one part of acid to nine parts water in a nonmetallic container.   Completed ☐

> **WARNING!**
> Pour the acid into the water, not the water into the acid! Be careful with this chemical and wear eye and skin protection.

6. Use a long-handled fiber brush to scrub the wall.   Completed ☐

7. Keep the area not being cleaned flushed free of acid and dissolved mortar. This scum, if allowed to dry, may beimpossible to remove later.   Completed ☐

8. Scrub the bricks, not the mortar joints. Do not use metal tools. Clean only a small area at a time.   Completed ☐

9. Rinse the wall thoroughly with plenty of clean water while the wall is still wet from scrubbing with the acid.   Completed ☐

Instructor's Initials: _____

Date: _____

Job 31 Review

After completing this job successfully, answer the following questions.

1. How long should you usually wait to see the effect of using a cleaning solution on a brick wall?

   _____
   _____

2. What acid is frequently used to clean brick walls?

   _____

3. What personal safety precautions should be taken when cleaning a brick wall with acid?

   _____
   _____
   _____

4. What are the usual mixing proportions of muriatic acid to water?

   _____

5. What precautions should be taken when mixing acid and water together?

   _____
   _____
   _____

6. What should you do before scrubbing a brick wall with acid?

   _____
   _____
   _____

7. Why shouldn't you scrub mortar joints with acid?

   _____
   _____
   _____

Score: _____

Name _____   Date _____   Class _____

JOB 32: Acid Cleaning of Light Colored Brick

OBJECTIVE: After completing this job, you will be able to clean new, light colored brick masonry with an acid solution using the proper technique.

TEXTBOOK REFERENCE: Study the appropriate section in Chapter 12, *Laying Brick*, before starting this job.

Equipment

To complete this job, you will need the following tools and materials:

- Eye and skin protect
- Long-handled stiff-fiber brush
- Mixing bucket (plastic)
- Wooden paddle or scraper
- Hydrochloric (muriatic) acid
- Potassium hydroxide or sodium hydroxide

Recommended Procedure

1. Remove large particles of mortar with wooden paddles or scrapers before wetting the wall. A chisel or wire brush might be necessary.   Completed ☐

2. Saturate the wall with clean water and flush away all loose mortar and dirt.   Completed ☐

> **WARNING!**
> When mixing the acid solution, pour the acid into the water, not the water into the acid! Be careful with this chemical and wear eye and skin protection.

3. Mix one part acid with 15 parts water.   Completed ☐

> **TRADE TIP**
> When cleaning light colored brick, use the highest-grade acid available. It should be free of any yellow or brown coloration.

4. Scrub the wall with a fiber brush.   Completed ☐
5. Rinse the wall well with clear water.   Completed ☐
6. Neutralize the acid wash with a solution of potassium hydroxide or sodium hydroxide, consisting of 1/2 lb hydroxide to 1 qt of water (2 lb/gal).   Completed ☐
7. Allow the acid solution to remain on the wall for two or three days before washing again with clear water.   Completed ☐

Instructor's Initials: _____

Date: _____

Job 32 Review

After completing this job successfully, answer the following questions.

1. When should you begin the cleaning operation?

   _____
   _____
   _____

2. What should be used to neutralize the acid wash?

   _____
   _____
   _____

3. Does the quality of the acid used to clean the brick affect the results? What should you look for when choosing an acid for cleaning light colored brick?

   _____
   _____
   _____

   Score: _____

Name _____ Date _____ Class _____

JOB 33 — Handling Concrete Blocks

OBJECTIVE: After completing this job, you will be able to handle concrete blocks using the proper technique.

TEXTBOOK REFERENCE: Study the appropriate section in Chapter 13, *Laying Block,* before starting this job.

Equipment

To complete this job, you will need the following tools and materials:

- Mason's tools
- Mortar
- Several 8″ × 8″ × 16″ concrete blocks

Recommended Procedure

Concrete blocks are large units that generally require both hands for placement. For example, a typical 8″ × 8″ × 16″ concrete block made with sand and gravel aggregates weighs 22 to 28 lb, depending on the specific aggregate used. Review Job 10 before starting this job.

1. Form a mortar bed for concrete block. Place one block (no head joint) on the mortar bed. Use both hands to lift the block, **Figure 33-1**. Grasp the web at each end of the block to lay it on the mortar bed. Even movements are preferred to jerking motions.

Author's image taken at Job Corps, Denison, IA

Figure 33-1. Placing a concrete block on the mortar bed.

2. The trowel should remain in your hand while placing the block to save time when only one or two blocks are set. If several blocks have been prepared, then lay the trowel aside while the blocks are placed on the mortar bed. Lift the block slowly and move it to the location where it is to be placed.   Completed ☐

3. By tipping the block slightly forward toward you when you place it on the mortar bed and looking down the face of the block, you can position the block in proper position with respect to the top edge of the course below. See **Figure 33-2**.

Completed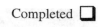

Author's image taken at Job Corps, Denison, IA

Figure 33-2. Placing a block in the wall.

4. Roll the block back slightly to correctly align the top of the block with the line. During this movement, the block should be pressed toward the last block to form a good head joint. Mortar should squeeze out slightly.

    Completed ☐

5. Blocks are positioned in a wall with the wide flange on the top. This provides a wider space for the bed joint. As each block is laid, cut off the excess mortar with the trowel held at a slight angle to the block.

    Completed ☐

6. Practice setting several blocks on the mortar bed using the procedure described in steps 2 to 5.

    Completed ☐

7. Clean up the work area, remove mortar from the floor and blocks, and return all materials and tools to their assigned places.

    Completed ☐

Instructor's Initials: _____

Date: _____

Name _____

Job 33 Review

After completing this job successfully, answer the following questions.

1. What is the approximate weight of a typical concrete block (8″ × 8″ × 16″)?
   _____

2. Describe the recommended method of grasping a concrete block.
   _____
   _____
   _____

3. Why are smooth, even movements recommended when placing a concrete block on the mortar bed?
   _____
   _____
   _____

4. Why should you tip the concrete block forward when you are placing it in the wall?
   _____
   _____
   _____

5. Is the wide flange of the block placed facing up or down? Why?
   _____
   _____
   _____

Score: _____

Notes

Name _____ Date _____ Class _____

JOB 34: Laying an 8″ Running Bond Concrete Block Wall

OBJECTIVE: After completing this job, you will be able to lay an 8″ running bond concrete block wall using the proper technique.

TEXTBOOK REFERENCE: Study the appropriate section of Chapter 13, *Laying Block,* before starting this job.

Equipment

To complete this job, you will need the following tools and materials:

- Mason's tools
- Chalk line
- Mortar
- Supply of blocks

Recommended Procedure

A well-planned concrete block structure involves mainly stretcher and corner blocks. See **Figure 34-1**. These blocks are nominally 8″ × 8″ × 16″. Actual size is 7 5/8″ × 7 5/8″ × 15 5/8″, allowing for a standard 3/8″ mortar joint.

Concrete blocks must be protected from excess moisture before use. If they are wet when placed, they will shrink when dry and cause cracks. Protect the blocks from rain.

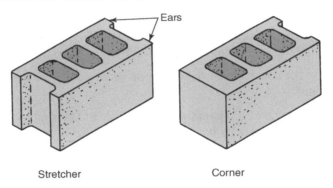

Goodheart-Willcox Publisher

Figure 34-1. These are the two primary concrete blocks used.

1. Collect the materials and tools needed and set up a work area. Establish the outside wall line with a chalk line. This line should be checked for squareness and proper length before proceeding to the next step. Ask your instructor what length of wall to build. This job procedure follows the photographic sequence in Chapter 13 of the text. Completed ☐

2. String out the block for the first course without mortar to check the layout. Allow 3/8″ for each mortar joint. Set the blocks aside when you are satisfied with the layout. Completed ☐

3. Spread a full mortar bed and furrow it with the trowel. Provide plenty of mortar on which to set the blocks. Completed ☐

4. Lay the corner block, positioning it carefully and accurately. Concrete blocks should be laid with the thicker edge of the face shell up to provide a wider mortar bed. Completed ☐

5. Lay several stretcher blocks along the wall line. Several blocks can be buttered on the end of the face shells if they are stood on end. This speeds the operation. To place them, push them downward into the mortar bed and sideways against the previously laid block. Completed ☐

6. After three or four blocks have been placed into position, they may be aligned, leveled, and plumbed with the mason's level. Tap on the block rather than the level. Completed ☐

7. After the first course has been laid, the corner lead is built up as shown in **Figure 34-2**. This corner is important, since the remainder of the wall is dependent upon its accuracy. The lead corner is usually laid up to four or five courses high above the center of the wall. Each course is checked as it is laid to be sure it is aligned, level, and plumb. Completed ☐

Figure 34-2. A typical concrete block lead.

Author's image taken at Job Corps, Denison, IA

8. Construct the other lead so a mason's line can be used to indicate the proper height of each course between the corners. Use line blocks, line pins, or adjustable line holders. Work from the corners toward the center of the wall. Do not allow a block to touch the line. Completed ☐

9. Complete the wall by laying to the line. Keep each block a line width away from the line and even at the bottom of the course. Completed ☐

10. Tool the mortar joints with the concave or V-joint tool. The tool should be slightly larger than the width of the mortar joint. A 5/8″ diameter bar is usually used for a 3/8″ concave mortar joint, and a 1/2″ square bar is used for making a 3/8″ V-shaped joint. Tools for tooling horizontal mortar joints should be at least 22″ long. Remove tags with the trowel. Completed ☐

11. When the mortar is sufficiently dry, the wall may be brushed or rubbed with a stiff-fiber brush or a burlap bag to remove dried particles. Completed ☐

12. Inspect your work, clean up the area, and return tools and materials to their assigned places. Completed ☐

Instructor's Initials: _____

Date: _____

Job 34 Review

After completing this job successfully, answer the following questions.

1. What two types of blocks are most well-planned concrete block walls composed of?

_____

_____

Name _____

2. What is the standard mortar joint thickness used with concrete blocks?

   _____

3. What is the actual size of an 8″ × 8″ × 16″ concrete block?

   _____

4. Why should concrete blocks be dry before they are laid?

   _____

   _____

   _____

5. What kind of bed is laid down for the first course of concrete blocks in a wall?

   _____

   _____

6. Why is the height of each course so critical in a lead?

   _____

   _____

   _____

7. How high is the lead corner generally in a concrete block wall?

   _____

                               Score: _____

Notes

Name _____ Date _____ Class _____

JOB 35 / Laying a 10" Concrete Block Cavity Wall

OBJECTIVE: After completing this job, you will be able to lay a 10" concrete block cavity wall using proper technique.

TEXTBOOK REFERENCE: Study the appropriate section of Chapter 13, *Laying Block*, before starting this job.

Equipment
To complete this job, you will need the following tools and materials:

- Mason's tools
- Mortar
- Supply of blocks
- Rectangular metal ties
- Flashing
- Weep hole wicking
- Rigid foam insulation (optional)
- Board to catch mortar droppings

Recommended Procedure
A cavity wall consists of two walls (wythes) separated by a continuous air space 2" to 4 1/2" wide. The wythes are tied securely together with noncorroding metal ties that are embedded in the mortar joints. See **Figure 35-1**. Unit ties are generally placed at every other block horizontally and in every other horizontal joint. Continuous metal joint reinforcement could be used instead.

Weep holes and flashing at the bottom of the wall need special attention if the wall is to be water resistant. Weep holes should be located in the bottom course at about every second or third head joint in the outside wythe. The cavity must be kept free of mortar droppings that could form a bridge between the wythes.

Portland Cement Association

Figure 35-1. Rectangular ties used in cavity wall construction.

1. Set up the work area and stock it with a good supply of 4" × 8" × 16" concrete blocks. Lay down a full bed of mortar for the inside wythe. This bed must be watertight.   Completed ☐

2. Place the blocks for the first course of the inside wythe. Be sure all head joints are solid and watertight.   Completed ☐

3. Position the flashing over the top edge of the first inside course and rest it on the foundation (under the outside wythe). Embed the flashing with mortar between the first and second course of the inside wythe.   Completed ☐

4. Lay the first course of the outside wythe allowing 2″ continuous air space between the wythes. The flashing will be beneath this course. Maintain weep holes between every second or third block. Insert wick or other acceptable material. Keep the cavity clean as the work proceeds up to the second course.   Completed ☐

5. Lay the second outer course with full bed and head joints. Be sure all joints and holes are filled with mortar. Do not allow any mortar to fall into the cavity. Mortar can be spread about 1/2″ back from the edge of the cavity to reduce the chance of its falling into the cavity. If rigid insulation is to be used in the wall, now is the time to install it. It should be placed against the inside wythe and should not prevent the escape of moisture from the cavity. A 1″ air space should always be maintained as the minimum.   Completed ☐

6. Position rectangular unit ties over the second course of the two wythes. Be sure they are the same height. Embed the ties within the mortar joint under the third course on the inside wythe.   Completed ☐

7. Lay the third outer course with the reinforcement embedded in the mortar joint.   Completed ☐

8. Place a board on the ties to catch mortar droppings as the next course is laid. The board will be removed when the next reinforcement is placed.   Completed ☐

9. Repeat the procedure until the wall has reached the desired height.   Completed ☐

10. Inspect your work, clean up the area, and return all tools and materials to their proper places.   Completed ☐

Instructor's Initials: _____

Date: _____

Job 35 Review

After completing this job successfully, answer the following questions.

1. In general, how often are metal ties placed in a 10″ concrete block cavity wall?

   _____

   _____

2. What provision for the escape of moisture did you make in your cavity wall?

   _____

   _____

   _____

3. How was the flashing positioned in the wall?

   _____

   _____

   _____

4. Why are full bed and head joints important in cavity wall construction?

   _____

   _____

Name _____

5. If rigid foam insulation is placed inside the cavity, what is the minimum air space?

   _____

6. What technique is recommended to catch mortar droppings in the cavity?

   _____

   _____

   _____

   Score: _____

Notes

Name _____ Date _____ Class _____

JOB 36: Laying an 8" Composite Wall with Concrete Block Backup

OBJECTIVE: After completing this job, you will be able to lay an 8" composite wall with concrete block backup using the proper technique.

TEXTBOOK REFERENCE: Study the appropriate section in Chapter 13, *Laying Block*, before starting this job.

Equipment

To complete this job, you will need the following tools and materials:

- Mason's tools
- Chalk line
- Mortar
- Supply of 4" × 8" × 16" concrete blocks
- Supply of common bricks
- Z ties
- Plasterer's trowel

Recommended Procedure

A composite wall is two wythes bonded together with masonry, metal ties, or joint reinforcement. The two wythes are joined together in a continuous mass using a vertical collar joint that prevents the passage of water through the wall.

Concrete blocks are often used as backup for bricks to make a composite wall. You will build this type of wall in this job. See **Figure 36-1**.

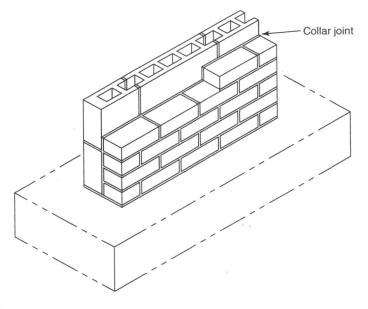

Goodheart-Willcox Publisher

Figure 36-1. 8" composite wall with brick facing and concrete block backup.

1. Arrange your workspace and stock it with a good supply of 4" × 8" × 16" concrete blocks and common bricks. Plan to build a wall that is 32" long and 32" high.   Completed ☐

2. Snap a chalk line on the floor to preserve the location of the wall. Cut several bricks in half and obtain six half blocks or cut your own.   Completed ☐

Copyright Goodheart-Willcox Co., Inc.
May not be reproduced or posted to a publicly accessible website.

3. Lay down a mortar bed for the inner wythe of 4″ concrete blocks. Lay a corner block at either end with a half block between them. Completed ☐

4. Lay the second course of the inner wythe of 4″ concrete blocks. Stagger the mortar joints. Parge the side adjacent to the outer wythe, being careful not to upset the bond. Use a plasterer's trowel because it is more efficient for this operation than a brick trowel. Completed ☐

5. Lay six courses of bricks in running bond to bring the outer wythe to 16″ high. Refer to **Figure 36-1**. Be sure the vertical collar joint is filled with mortar and the bricks are laid with full bed joints. Completed ☐

6. Position two Z ties across the wythes about 24″ apart. Embed them in mortar. Completed ☐

7. Bring the wall up another course of concrete blocks to produce a height of 32″. Complete the outer wythe of bricks. Completed ☐

8. Strike the joints and brush the completed wall. Completed ☐

9. Inspect your work, clean up the area, and return all tools and materials to their assigned places. Completed ☐

Instructor's Initials: _____

Date: _____

Job 36 Review

After completing this job successfully, answer the following questions.

1. What is the function of the collar joint in a composite wall?

   _____
   _____
   _____

2. What is the proper name for 4″ × 8″ × 16″ concrete blocks?

   _____

3. What tool was recommended for parging the blocks on the inner wythe adjacent to the outer wythe?

   _____

4. How far apart were the Z ties placed?

   _____

5. What tool was used to strike the horizontal joints?

   _____

Score: _____

Name _____ Date _____ Class _____

JOB 37: Cleaning Concrete Block Masonry

OBJECTIVE: After completing this job, you will be able to clean concrete block masonry using proper technique.

TEXTBOOK REFERENCE: Study the appropriate section in Chapter 13, *Laying Block,* before starting this job.

Equipment

To complete this job, you will need the following tools and materials:
- Eye and hand protection
- Mason's trowel
- Chisel or putty knife
- Piece of concrete block
- Commercial cleaning agent

Recommended Procedure

Concrete block walls are not cleaned with acid to remove mortar smears or droppings. Therefore, care must be taken to keep the wall surface clean during construction.

1. Allow any mortar droppings that stick to the wall to dry and harden before attempting to remove them.   Completed ☐

2. Remove large particles of dry mortar with a trowel, chisel, or putty knife. If you attempt to remove wet mortar, it will most likely smear into the surface of the block and become permanent.   Completed ☐

3. Rub the wall with a piece of concrete block to remove practically all of the mortar.   Completed ☐

4. If further cleaning is necessary, you can use a commercial cleaning agent, such as a detergent. Be sure to follow the manufacturer's directions and try the product on a small section of the wall to check the results.   Completed ☐

Instructor's Initials: _____

Date: _____

Job 37 Review

After completing this job successfully, answer the following questions.

1. Why do you think concrete block walls are not cleaned with acid?

   _____

   _____

   _____

2. When should you try to remove any mortar droppings that stick to the wall?

   _____

   _____

3. How did you remove large particles from the wall?

_____
_____
_____

4. After removing hardened particles from the wall, what is the next step?

_____
_____
_____

5. If a wall still has smears after the recommended cleaning procedure has been followed, then what can you do?

_____
_____
_____

Score: _____

Name _____ Date _____ Class _____

JOB 38 — Handling Stone

OBJECTIVE: After completing this job, you will be able to handle stone using the proper technique.

TEXTBOOK REFERENCE: Study the appropriate section of Chapter 14, *Stonemasonry,* before starting this job.

Equipment
To complete this job, you will need the following tools and materials:
- Safety glasses or goggles
- Steel-toed shoes and leather gloves
- Pry bar
- Mason's hammer
- Sledgehammer
- Chisel
- Variety of stone samples

Recommended Procedure
This job is divided into two sections. The first section pertains to handling field stone. The second section pertains to handling ashlar, trimmings, and panels.

Field Stone
Stone used for rubble or roughly squared stonework will most likely be field stone in the sizes and shapes as they are found in fields and streams. The stones in a load will vary as to type. A skilled stonemason knows which stones can be split easily and which shapes will look best in various locations in the wall. Learn from your own experience and work with a professional to discover the tricks of the trade. For example, bedding is visible in most sedimentary stones, and an experienced mason will know just where to strike the stone to split it.

1. Keep the pile of stone close to where you are working, but clear of the work area. If the source of stone is too far away, you will spend most of your time running back and forth. If it is too close, you will not have any room to work and may fall over the stones. Spread out the stones and look them over to see what you have.   Completed ☐

2. Form a particular pattern in your mind before you begin setting stone. The pattern should take into consideration the size, shape, and type of stone you have to work with. If all the stones are round and smooth, do not plan a pattern that requires cutting and trimming every stone. If you plan to build a polygonal stone pattern, then order broken stone that already has the basic shapes needed. Study the basic patterns shown in **Figure 38-1**.

Completed ☐

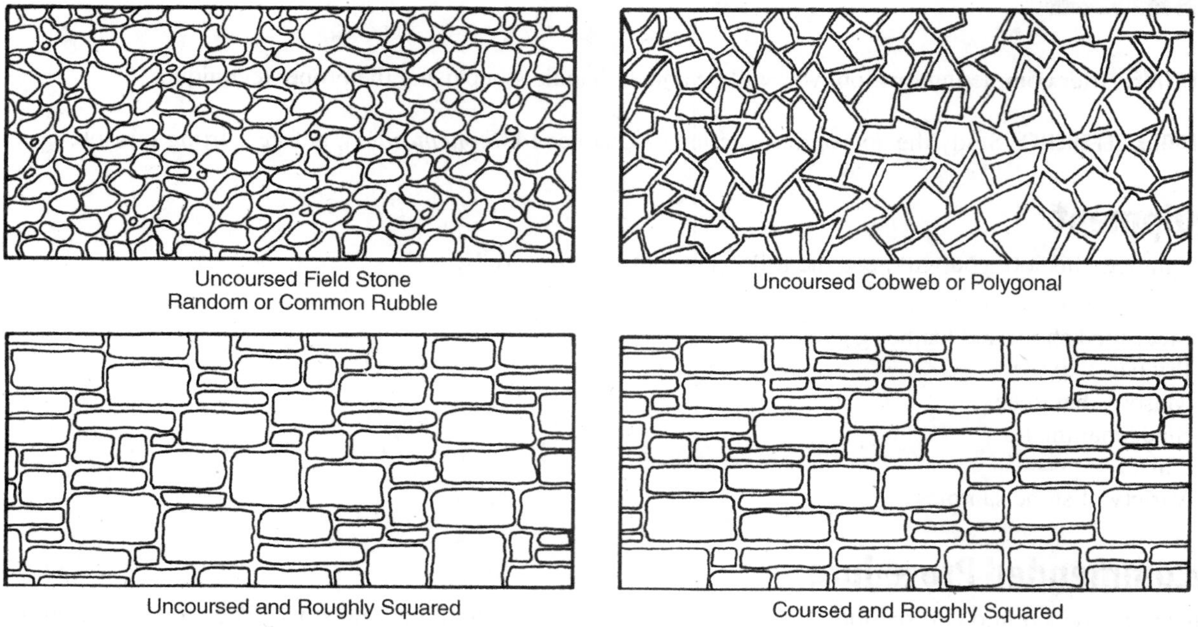

Uncoursed Field Stone
Random or Common Rubble

Uncoursed Cobweb or Polygonal

Uncoursed and Roughly Squared

Coursed and Roughly Squared

Goodheart-Willcox Publisher

Figure 38-1. Representative stone patterns that use field stones.

3. Pick out a large stone to break into smaller pieces for an uncoursed and roughly squared stone pattern. Do not try to lift large stones by yourself. Remember that a cubic foot of granite weighs about 170 lb. Use pry bars to move large stones.

Completed ☐

WARNING!
Wear protective clothing when working with stone—safety glasses or goggles, steel-toed shoes, and heavy leather gloves.

4. Examine the stone that has been selected for any potential break lines. Try to visualize where the stone might break. Strike it with your mason's hammer to see the results. If the stone does not break, use a larger hammer or sledgehammer.

Completed ☐

5. Next, try to remove an unwanted protuberance on a flat side of a stone with a chisel. Crush or powder a point on a stone with the mason's hammer. The purpose of this job is to get the feel of working with rubble stone.

Completed ☐

Ashlar, Trimmings, and Panels

Cut stone in the form of ashlar, trimmings, or panels must be handled carefully to prevent breakage, staining, and chipping. This stone is delivered to the jobsite already cut, dressed, and finished to precise specifications for a particular job. See **Figure 38-2**.

Name _____

1. Examine several pieces of cut stone to determine the condition of each piece. Make note of chips, scratches, broken pieces, or discolored stone. Completed ☐

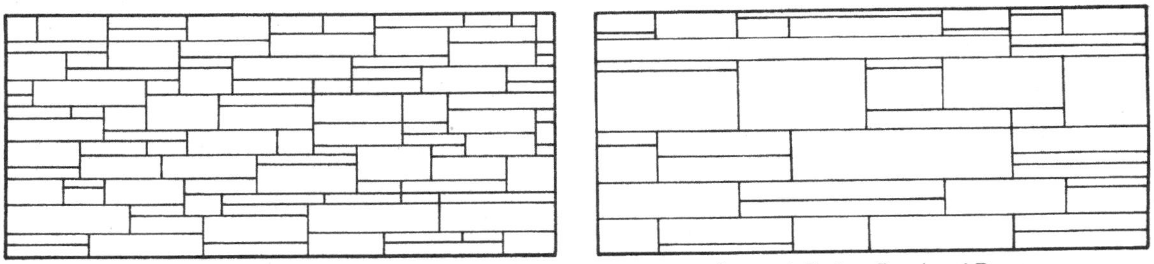

Random, Broken Course and Range

Coursed, Broken Bond and Range

Goodheart-Willcox Publisher

Figure 38-2. Ashlar stone patterns.

2. Handle this stone as you would any other material that you do not want to chip or scratch. Do not slide it on a rough surface. Store it on sturdy skids or timbers in a dry place. Completed ☐

3. Lean smooth-finished stones face-to-face and back-to-back. Textured finishes should be separated with spacers. Be careful not to bump stones. Completed ☐

4. Examine large stone panels to see if they have lifting holes. Discuss safety procedures for lifting large stone panels with your instructor. Estimate the weight of a large panel by figuring its volume and weight per cubic foot. Completed ☐

5. Become familiar with the various types of stone by name and application. Get to know your material. Completed ☐

Instructor's Initials: _____

Date: _____

Job 38 Review

After completing this job successfully, answer the following questions.

1. What kind of stone is generally used for rubble or roughly squared stonework?
   _____

2. In what type of stone is bedding usually visible?
   _____

3. What is the problem with keeping your source of stone too close to where you are building the wall?
   _____
   _____
   _____

4. Once the stone is delivered, what should you do first?
   _____
   _____
   _____

5. How did you remove small knobs or protuberances from stones?

_____
_____
_____

6. How should textured panels be stacked?

_____
_____
_____

Score: _____

Name _____ Date _____ Class _____

JOB 39 / Forming Mortar Joints in Stone Masonry

OBJECTIVE: After completing this job, you will be able to form mortar joints in stone masonry using the proper technique.

TEXTBOOK REFERENCE: Study the appropriate section of Chapter 14, *Stonemasonry*, before starting this job.

Equipment

To complete this job, you will need the following tools and materials:
- Stonemason's tools
- Mortar
- Wood wedges
- Assortment of broken stone pieces
- Assortment of cut stone pieces
- Narrow caulking trowel

Recommended Procedure

Whenever mortar is used with stone, it should be a nonstaining type designed for the specific application. A mix of one part nonstaining cement, one part hydrated lime, and six parts clean, sharp, washed sand is recommended.

The width of the mortar joints affects the finished appearance of the stone construction. The most frequent error is to allow the mortar joints to become too wide, especially with the use of rubble or polygonal stones. Stones should be cut to fit with mortar joints that are 1/2″ for rough work and 3/8″ for ashlar. Use a narrow caulking trowel for filling narrow spaces and working mortar into crevices.

In this job you will form mortar joints using polygonal stone. See **Figure 39-1** to visualize the basic layout to be attempted in this job.

Goodheart-Willcox Publisher

Figure 39-1. Polygonal stonework.

1. Select a vertical masonry or wood backing about 24″ × 36″ in size. You will lay your stonework against it. Completed ☐

2. Select an assortment of polygonal-shaped stones of various sizes, but not over 1 sq ft in area. Completed ☐

3. Arrange the stones in a pleasing pattern on the floor. You may have to chip away some part of a stone to make it fit, but try to find pieces that naturally fit together. Completed ☐

4. When you have selected several pieces that will cover about 4 sq ft, mix some mortar (not too thin) and lay a mortar bed long enough for two or three of the larger stones. Larger stones are generally placed at the bottom of a wall. Completed ☐

5. Set the base stones in the mortar bed and against the backup. If the stones are not stable, or too much mortar squeezes out, place one or two wood wedges under each stone. Remember to keep mortar joints narrow. Completed ☐

6. Continue along the base until the width is completed. Remove excess mortar and try to keep it off the face of the stones. Completed ☐

7. Pick up the next stone to be placed and see how it fits the space intended for it. You may have to try several stones to get the right fit. Completed ☐

8. When you are sure the stone will fit properly, lay a bed of mortar on the stone already in place and set this stone. Remove excess mortar and insert wedges if needed. Completed ☐

9. Continue the process until you have covered the space. Be very careful around the structure, because it may fall easily. Completed ☐

10. When the mortar has hardened somewhat, carefully rake out some mortar between the stones until the joint is about 1″ deep. You may remove some of the wedges. If your work was very good, tool the joints with a short jointer. Carefully add mortar with the narrow caulking trowel where it is needed to replace the mortar removed earlier. Completed ☐

11. When the mortar has hardened, remove the wedges and brush the stone to remove mortar splatters. Fill any holes left when the wedges were removed. Completed ☐

12. Inspect your work to see how your mortar joints look. Clean up the area and return the tools and materials to their assigned places. Completed ☐

Instructor's Initials: _____

Date: _____

Job 39 Review

After completing this job successfully, answer the following questions.

1. What is the recommended mortar mix for stone masonry?

_____
_____
_____

2. What is the most frequent error in forming mortar joints in stone masonry?

_____
_____
_____

3. How did you prevent large stones from squeezing out too much mortar?

_____
_____
_____

4. What is the general approach to shaping polygonal stone?

_____
_____
_____

5. What tool is generally used to point stonework?

_____

Score: _____

Name _____ Date _____ Class _____

JOB 40: Splitting, Shaping, and Cutting Stone

OBJECTIVE: After completing this job, you will be able to split, shape, and cut stone for stonemasonry work using the proper technique.

TEXTBOOK REFERENCE: Study the appropriate section of Chapter 14, *Stonemasonry*, before starting this job.

Equipment

To complete this job, you will need the following tools and materials:

- Stonemason's tools
- Stonemason's hammer
- Sledgehammer
- Pry bar
- Steel wedges
- Electric drill (optional)
- Stonemason's chisel
- Safety glasses, steel-toed shoes or boots, and other protective gear

Recommended Procedure

Splitting stone takes practice and sharp observation to achieve the desired results in a consistent manner.

1. To split a stone that has a stratified (layered) structure, mark a line along the grain, then chip on the line with the chisel end of your mason's hammer until a crack begins to develop.   Completed ☐

2. Widen the crack slowly by driving wedges into the crack at several points. If the stone is very large, a pry bar may be needed to finish prying it apart. Force the stone apart.   Completed ☐

3. Select a large stone, such as granite, that is not stratified. These stones are difficult to split, but it is possible. Begin by drilling holes with the narrow-bladed chisel or power drill about 6″ apart along the line where you want the stone to split.   Completed ☐

4. Drive thin wedges into the holes. Continue the process until the stone splits.   Completed ☐

5. Select a split stone and shape it with your mason's hammer. Also try the stonemason's chisel. Be sure you are wearing safety glasses.   Completed ☐

6. Ask your instructor to demonstrate the use of the power masonry saw. It will be used to cut a piece of ashlar to a particular dimension. Never use any equipment until the instructor gives approval.   Completed ☐

> **WARNING!**
> Safety glasses or goggles are required when operating a power masonry saw. A face shield should also be worn to protect from flying chips. Ear protection is needed because of the amount of noise created by sawing.

7. Select a piece of cut stone and plan a cut across the end. Hold the stone firmly against the fence and slowly make a cut toward you. When the cut is finished, return the saw to the starting position. Examine your cut. Make another cut to get the feel for the saw.

Completed ☐

WARNING!
Never try to cut large panels or pieces too big for the saw.

Instructor's Initials: _____

Date: _____

Job 40 Review

After completing this job successfully, answer the following questions.

1. Briefly, what is the procedure for splitting a stone that has a stratified structure?

   _____
   _____
   _____
   _____

2. What is the recommended procedure for splitting a stone that is not stratified?

   _____
   _____
   _____
   _____

3. What personal safety equipment is recommended for working with stone?

   _____
   _____

4. What direction is the cut made when using the power masonry saw?

   _____

5. In what position should a piece of stone be held when cutting with the masonry saw?

   _____

Score: _____

Name _____ Date _____ Class _____

JOB 41: Setting a Random Rubble Stone Veneer Wall

OBJECTIVE: After completing this job, you will be able to set a random rubble stone veneer wall using the proper technique.

TEXTBOOK REFERENCE: Study the appropriate section in Chapter 14, *Stonemasonry*, before starting this job.

Equipment

To complete this job, you will need the following tools and materials:

- Stonemason's tools
- Mortar
- Safety glasses and other protective gear
- Existing wall masonry or frame
- Supply of rubble stone
- Corrugated ties
- Piece of chalk
- Bucket of water and sponge

Recommended Procedure

This job requires applying stone veneer to an existing wall. The backup wall may be masonry or frame, but it should already have, or provide for the attachment of, flexible ties or corrugated fasteners. The wall area should be at least 4′ wide by 4′ high. **Figure 41-1** shows what the completed wall should look like.

Goodheart-Willcox Publisher

Figure 41-1. Random rubble stonework with uniform mortar joints.

1. Arrange your work area and spread out the stones so each one can be examined for shape, texture, or other characteristics. The stones will most likely vary in size from 6″ to 18″.   Completed ☐

2. Mix enough mortar to last for one hour of work. Use it up before it begins to set. Be sure to use a nonstaining mortar. Mix one part nonstaining cement, one part hydrated lime, and six parts clean, sharp, washed sand. A comparable manufactured mortar cement can also be used.   Completed ☐

3. Select several larger stones for the bed course and place them into position dry. With a piece of chalk, mark them for trimming.   Completed ☐

Stones should be placed in as natural a position as possible. Have a pattern or special effect in mind as you select and place stones together. Do not use too many varieties and keep textures relatively uniform.

Stone veneer is generally 4″ to 8″ thick, depending on the method of bonding and construction specifications. See **Figure 41-2**.

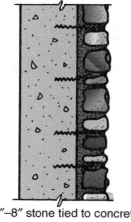

4″–8″ stone tied to concrete backing using wall ties.

4″–8″ stone tied to frame construction. Use wood sheathing and W.P. felt or W.P. sheathing board.

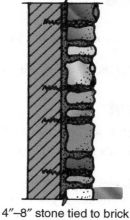

4″–8″ stone tied to brick or block masonry. Provide 1″ – 2″ air space or slush fill voids.

Goodheart-Willcox Publisher

Figure 41-2. Stone veneer sections.

Large stones must be split so they meet these requirements. A sledgehammer may be used for splitting large stones.

4. Trim each stone as you are ready to place it in the wall. Thoroughly clean each stone on all exposed surfaces by washing with a brush and soap powder, followed by a thorough drenching with clean water.   Completed ☐

> **WARNING!**
> Be careful of flying chips! Wear safety glasses.

5. Place each trimmed stone in its proper location in the wall and check for proper fit. When you are satisfied with the result, proceed to the next step.   Completed ☐

6. Using the trowel, lay a generous full bed of mortar for the trimmed stones and place them into position.   Completed ☐

7. Fill in the spaces between the stones with mortar. A narrow trowel works well for filling narrow spaces and working the mortar into crevices. See **Figure 41-3**. Joints should be about 1/2″ to 1″ for rough work such as this. Sponge the stone free of mortar along the joints as the work progresses.   Completed ☐

schankz/Shutterstock.com

Figure 41-3. A narrow trowel.

Name _____

8. Remove excess mortar from the stone with the trowel and strike the joint. Joints can be tooled when initial set has occurred. If desired, they can be raked out 1″ deeper and pointed later with mortar.   Completed ☐

9. Trim larger stones for the deeper bed course and fit them into place using the same procedure performed when placing smaller stones.   Completed ☐

10. Bend corrugated metal ties into place for successive courses of stone. All ties must be noncorrosive. Use extra ties at all corners and large stones when possible.   Completed ☐

Lead, plastic, or wood pads the thickness of mortar joints should be placed under heavy stones to avoid squeezing mortar out. See **Figure 41-4**. Remove the pads after the mortar has set, then fill the holes. Heavy stones or projecting courses should not be set until mortar in the courses below has hardened sufficiently to avoid squeezing.   Completed ☐

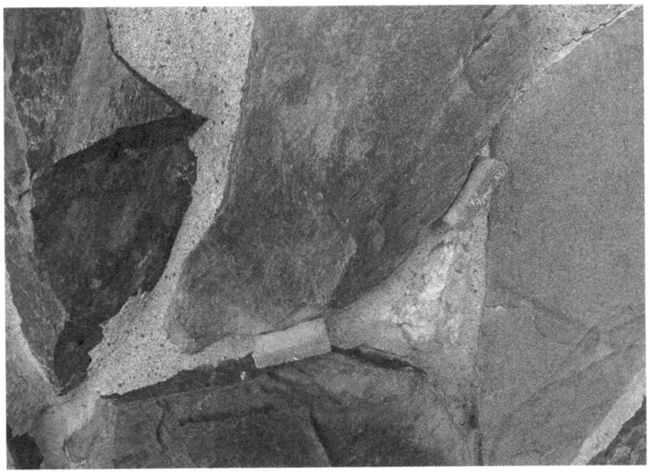

Timothy L. Andera

Figure 41-4. Wood pads supporting heavy stones.

11. After the mortar has set between the stones, continue laying new courses of stones until the wall is the desired height. Work from the corners toward the middle.   Completed ☐

A mason's line and level can be used to keep the wall straight and plumb. If the structure is inclined or tapered, as in many chimneys, a mason's line should be used to keep the edge straight and moving in the proper direction.

The masonry should be protected at all times from rain and masonry droppings. Adequate protection must be provided during cold-weather construction.

12. When the wall is completed, scrub it with a fiber brush and clean water. The stone should be clean and free of mortar.   Completed ☐

Waterproofing may be applied. Use a nonstaining asphalt emulsion, vinyl lacquer, cement base masonry waterproofing, stearate, or other approved material.

TRADE TIP
Strong acid compounds generally should not be applied, since they may burn and discolor certain types of stone, such as limestone.

13. Inspect your work, clean up the area, and return all tools and materials to their assigned places.   Completed ☐

Instructor's Initials: _____

Date: _____

Job 41 Review

After completing this job successfully, answer the following questions.

1. Any load of rubble stone is most likely composed of stones of many sizes. Is there a general rule to follow in placing larger stones?

   _____

   _____

   _____

2. Is coursing possible in a rubble stone masonry wall? Why?

   _____

   _____

   _____

3. How does the strength of mortar bonds in a rubble stone wall compare with other types of stone walls?

   _____

   _____

   _____

4. What is the overall appearance or style communicated by a rubble stone wall?

   _____

   _____

5. What type of metal ties are generally used to bond a rubble stone wall to its backup?

   _____

6. How much mortar should you mix when placing stone masonry?

   _____

   Score: _____

Name _____ Date _____ Class _____

JOB 42 — Building a Solid 12″ Thick Ashlar Stone Wall

OBJECTIVE: After completing this job, you will be able to build a solid ashlar stone wall using the proper technique.

TEXTBOOK REFERENCE: Study the appropriate section in Chapter 14, *Stonemasonry*, before beginning this job. Also review Job 41.

Equipment
To complete this job, you will need the following tools and materials:
- Stonemason's tools
- Mortar
- Safety equipment
- Supply of ashlar stone (variety of dimensions)
- Bucket of water and sponge

Recommended Procedure
In this job, you will build an ashlar stone wall section 12″ thick, 24″ high, and 60″ long. The material should include widths of 6″ and 12″ and various thicknesses and lengths. **Figure 42-1** shows the material.

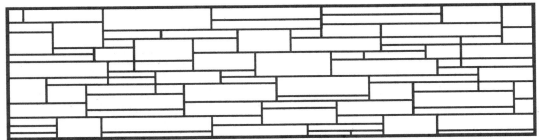

Goodheart-Willcox Publisher

Figure 42-1. Ashlar stone wall.

1. Collect the tools and materials needed for this job and arrange your work area. Be careful with the stone because it can chip easily. Completed ☐

2. Using a chalk line, snap a line to locate both sides of a wall 12″ thick and 60″ long. Be sure the ends are perpendicular to the face of the wall. Completed ☐

3. Lay out a pattern of stone on the floor to see how the various pieces can be assembled to provide the desired pattern. The lengths and thicknesses are usually designed to form a regular pattern with a standard mortar joint thickness. Completed ☐

4. When you have visualized the pattern, mix some mortar and lay a full mortar bed for the base course for the front wythe of the wall. Be sure the chalk line is visible. Completed ☐

5. Set the first stone at one end of the wall. Level, plumb, and align the stone to be sure it is properly located. Check the mortar joint thickness to be sure it is 3/8″. Completed ☐

6. Set the remaining stones on the face wythe using an interesting variety of sizes. Follow your planned pattern. Remove excess mortar, level, and straightedge the wall. Completed ☐

Copyright Goodheart-Willcox Co., Inc.
May not be reproduced or posted to a publicly accessible website.

7. Lay down a mortar bed for the back wythe of the wall. Be sure that some of the stones in this wythe are the same height as those opposite them so a bond stone (a stone that reaches across the total wall thickness) can be used at that point. Bond stones tie the wythes together and provide strength and stability. Plan to insert a bond stone at regular intervals throughout the wall. Check each stone to be sure it is level and plumb. Any irregularity will be very visible in the finished wall. Completed ☐

8. Continue building the wall and working your pattern. Be sure to use full mortar joints and try to avoid getting mortar on the face of the stone. Completed ☐

9. When your wall has reached 24″ high, cap it off using 12″ wide pieces all the same thickness. Select a thickness that looks right in proportion to the size of stone used in the wall and the wall's height and length. Completed ☐

10. Remove all excess mortar and tool the joints when the mortar has begun to harden. Remove mortar smears with a damp sponge. Rub the stone, not the mortar joints. Completed ☐

11. Inspect your work, clean up the area, and return your tools and materials to their assigned places. Completed ☐

Instructor's Initials: _____

Date: _____

Job 42 Review

After completing this job successfully, answer the following questions.

1. Define *ashlar stone*.

   _____
   _____
   _____

2. Is laying (setting) ashlar stone more similar to laying rubble stonework or laying bricks?

   _____
   _____
   _____

3. What lengths of ashlar stone are generally available?

   _____
   _____

4. What mortar thickness is commonly used with ashlar stone?

   _____

5. How many wythes did your 12″ thick ashlar stone wall have?

   _____

6. How were the wythes tied together?

   _____
   _____

Score: _____

Name _____ Date _____ Class _____

JOB 43: Setting a Limestone Panel

OBJECTIVE: After completing this job, you will be able to install a limestone panel using proper technique. *This is a class project.*

TEXTBOOK REFERENCE: Study the appropriate section in Chapter 14, *Stonemasonry,* before starting this job.

Equipment

To complete this job, you will need the following tools and materials:

- Stonemason's tools
- Mortar
- Typical safety equipment
- Limestone panels
- Lifting device and attachments
- Steel or concrete frame to receive the panels

Recommended Procedure

This job involves setting a limestone panel on a steel or concrete frame. The frame will be provided by the instructor. The Indiana Limestone Institute provides extensive procedures and criteria for the application of limestone panels. This job should be coordinated with its recommendations.

1. Study the construction drawings, which include details of the anchoring system. Examine the panel and make note of provisions for attachment to the structural frame. Completed ☐

2. Discuss with your instructor the method of lifting the stone panel into place. What safety precautions will be taken to protect you as well as to prevent damage to the panel? Completed ☐

3. Rehearse the process of lifting the panel and the roles to be played by various members of the class. Be sure you have all the tools and materials at hand to complete the job once it is begun. Completed ☐

> **WARNING!**
> Hard hats are required.

4. Lift the panel and carefully guide it to its desired placement. Check alignment of all support pins or other attachments. Check the panel to be sure it is level, plumb, and securely attached to the building frame. Completed ☐

5. Once the panel is securely anchored, review the process to discuss any problems or procedures that were unclear. Remove the panel and reinstall it so everyone has a chance to play the various roles in installing a large stone panel. Completed ☐

6. Return the panel to storage and put away the tools and equipment used in this job. Completed ☐

Instructor's Initials: _____

Date: _____

Job 43 Review

After completing this job successfully, answer the following questions.

1. How are limestone panels supported and attached to buildings?

   _____

   _____

   _____

2. How is a stone panel generally protected during the lifting and setting process?

   _____

   _____

   _____

3. How is setting a stone panel different from other stonework?

   _____

   _____

   _____

4. Why must a stone panel be handled very carefully?

   _____

   _____

   Score: _____

Name _____ Date _____ Class _____

JOB 44: Pointing Cut Stone after Setting

OBJECTIVE: After completing this job, you will be able to point cut stone using the proper technique.

TEXTBOOK REFERENCE: Study the appropriate section in Chapter 14, *Stonemasonry*, before starting this job.

Equipment

To complete this job, you will need the following tools and materials:

- Stonemason's tools
- Mortar
- Slicker
- Rake out jointer or skate wheel joint raker
- Stone assembly with mortar removed to 1″ deep

Recommended Procedure

Pointing cut stone after setting, rather than full bed setting and finishing in one operation, reduces a condition which tends to produce spalling and leakage. Pointing can be done in one, two, or three stages. This allows each stage to seal shrinkage cracks in the preceding stage, and the concave tooled joint provides the maximum protection against leakage.

1. When the mortar has hardened sufficiently in a mortar joint, use a skate wheel joint raker to remove the mortar to a depth of 1″. See **Figure 44-1**. Completed ☐

Stanley Goldblatt

Figure 44-1. The skate wheel joint raker.

2. Clean out the area where the mortar has been removed with the skate wheel joint raker. Use a brush or compressed air. The joint area must be clean of debris. Completed ☐

3. Mix a small quantity of mortar and apply about a 1/2″ thick layer of mortar to the existing joint. Use a slicker to apply the mortar between the pieces of cut stone. Press the mortar into place so it makes a good bond. Completed ☐

4. When the previous application of mortar has hardened, then the next (in this case, the last) application may be started. Fill the mortar joint with new mortar, again using the slicker or narrow pointing trowel. The mortar should be flush with the surface of the stone units. Completed ☐

5. When the mortar is thumbprint hard, tool the joints. Remove tags and brush the stone to remove any mortar droppings. Sponge the stone to further clean the surface. Do not sponge the mortar joints. Completed ☐

6. A commercial sealer may be applied, if desired. Follow the manufacturer's instructions. Completed ☐

7. Inspect your work, clean up the area, and return tools and materials to their assigned places. Completed ☐

Instructor's Initials: _____

Date: _____

Job 44 Review

After completing this job successfully, answer the following questions.

1. What is the purpose of pointing cut stone?

   _____
   _____
   _____

2. How deep is the mortar generally raked out for pointing?

   _____

3. How many stages are used to point cut stone after setting?

   _____

4. What tool is most often used to rake out the mortar for pointing?

   _____

5. What tool is used to add new mortar to the joint?

   _____

6. What is the purpose of applying a commercial sealer to cut stone?

   _____
   _____
   _____

Score: _____

Name _____ Date _____ Class _____

JOB 45: Cleaning New Stone Masonry

OBJECTIVE: After completing this job, you will be able to clean new stone masonry using the proper technique.

TEXTBOOK REFERENCE: Study the appropriate section in Chapter 14, *Stonemasonry,* before starting this job.

Equipment

To complete this job, you will need the following tools and materials:

- Eye and skin protection
- Stiff fiber brush
- Bucket and water
- Detergent
- Rinsing water (under pressure)

Recommended Procedure

The appearance of a masonry structure may be ruined by improper cleaning. In many instances, the damage caused by faulty cleaning techniques or use of the wrong cleaning agent cannot be repaired. All cleaning should be applied to a sample test area of approximately 20 sq ft.

Always take precautions to protect the stone from water, dirt, and mortar splatter as the work progresses. Cover the wall with a waterproof membrane at the end of each workday. Practice careful workmanship to prevent excessive mortar droppings.

Acids, wire brushes, and sandblasting are usually not permitted on stonework. Strong acid compounds used for cleaning brick burn and discolor many types of stone. Machine cleaning processes should be approved by the supplier before use on stone.

1. When the mortar is cured, brush off the stone with a stiff-fiber brush to remove any loose dirt or mortar. Cleaning uncured mortar can cause damage and spread scum over the stone surfaces.   Completed ☐

2. With a stiff fiber brush and clean water, scrub the wall. Concentrate on the stone, not on the mortar joints.   Completed ☐

3. If stains are difficult to remove, use soapy water and then rinse with clear water. If the stonework has been kept clean by sponging during construction, the final cleaning will be much easier.   Completed ☐

4. Examine the wall to see if any spots have been missed. Clean these spots.   Completed ☐

5. Return tools and materials to their assigned places.   Completed ☐

Instructor's Initials: _____

Date: _____

Job 45 Review

After completing this job successfully, answer the following questions.

1. What kind of personal safety protection is recommended when cleaning new stone masonry?

   _____

   _____

   _____

2. How can improper cleaning ruin a stone masonry job?

   _____

   _____

   _____

3. Test-cleaning an obscure area of stone is recommended to be sure the cleaning agent is compatible with the stone. How large an area is recommended, and how long do you wait to see the results?

   _____

   _____

4. What cleaning processes are generally not permitted on stone?

   _____

   _____

5. What is the possible effect of cleaning uncured mortar?

   _____

   _____

   _____

   Score: _____

Name _____ Date _____ Class _____

JOB 46 / Applying Manufactured Stone to a Backup

OBJECTIVE: After completing this job, you will be able to apply manufactured stone to a backup using the proper technique.

TEXTBOOK REFERENCE: Study the appropriate section in Chapter 9, *Stone,* before starting this job.

Equipment

To complete this job, you will need the following tools and materials:
- Plasterer's trowel
- Mason's trowel
- Slicker or narrow pointing trowel
- Stiff bristle brush
- Hatchet
- Grout bag

Recommended Procedure

Manufactured stone may be applied directly to a base coat of stucco, concrete block, brick, concrete, or any masonry surface that has not been treated or sealed and which is rough enough to provide a good mechanical bond. **Figure 46-1** shows a typical application of manufactured stone.

Stone Products Corporation

Figure 46-1. Applying manufactured stone to a masonry wall.

1. Prepare the surface. If the surface is clean (unpainted, unsealed, or untreated) brick, block, or concrete, the stone may be applied directly to the surface. On all other surfaces, metal lath must be applied first. Nail or staple the lath 6″ O.C. Use a vapor barrier (such as building felt) under the lath on outside applications.

Completed ☐

2. Apply the scratch coat. A thin scratch coat of mortar is generally applied and allowed to set prior to installation of the stone. Mortar for this operation can be mixed by combining masonry cement and sand or regular Portland cement, lime, and sand. The consistency should not be too thin or too dry. This mortar can be used for the scratch coat, for applying the stone, and for grouting the joints. Apply the scratch coat with a plasterer's trowel. — Completed ☐

3. Apply mortar to the stone. Lay out the stone near the work area and become familiar with the various shapes and sizes. When you are ready to begin, select a stone and apply a 1/2″ thick, even layer of mortar to the back of the stone. Use the mason's trowel for this operation. — Completed ☐

4. Apply the stone. Press the stone firmly into place on the wall surface so the mortar behind the stone squeezes out around all sides. Use a gentle wiggling action while applying the stone to ensure a good bond. Application of manufactured stone usually begins at the top to help keep the stone clean during construction. Install corner stones first to make fitting easier. Keep the mortar joints tight and uniform. — Completed ☐

5. Trim the stone. When necessary, manufactured stone can be cut and shaped with a hatchet, brick trowel, or nippers to form special sizes and shapes for a better fit. Try to position stones on the wall so cut edges will not show. Trim a stone to fit a particular spot. — Completed ☐

6. Grout the joints. After all the stone has been applied to the surface, fill a grout bag with mortar. Partially fill the joints between the stones with mortar. Be sure to cover broken stone edges with mortar. — Completed ☐

7. Strike the joints. When the mortar joints have hardened sufficiently, use a wood or metal striking tool to rake out the excess mortar to the desired depth and at the same time to force the mortar into the joints to seal the joint edges. — Completed ☐

8. Brush. Brush the mortar joints with a stiff bristle brush to smooth them and clean away any loose mortar. Remove mortar spots from the face of the stone. — Completed ☐

9. Waterproofing. Apply a high-quality waterproofing sealer to the surface. This will help keep the surface clean. — Completed ☐

10. Inspect your work, clean up the area, and return all tools and materials to their assigned places. — Completed ☐

Instructor's Initials: _____

Date: _____

Job 46 Review

After completing this job successfully, answer the following questions.

1. What kind of concrete is used to make manufactured stone?

_____

2. What kinds of bases are suitable for the application of manufactured stone?

_____

_____

3. What is a *scratch coat*?

_____

_____

_____

4. What tool is used to apply a scratch coat?

_____

Name _____

5. What mason's tool may be used to cut manufactured stone?

   _____

6. When applying manufactured stone, should you start at the top or bottom of the wall? Why?

   _____

   _____

   _____

   _____

                                    Score: _____

Notes

Name _____ Date _____ Class _____

JOB 47: Measuring Concrete Materials

OBJECTIVE: After completing this job, you will be able to measure concrete materials using the proper technique.

TEXTBOOK REFERENCE: Study the appropriate section in Chapter 18, *Concrete Materials and Applications,* before starting this job.

Equipment

To complete this job, you will need the following tools and materials:

- Scale to weigh materials
- One cubic foot box
- Dry and moist sand
- Unopened bag of cement
- Eye and hand protection

Recommended Procedure

The ingredients in each batch of concrete need to be accurately measured to produce uniform batches of proper proportions and consistency. The problem of varying amounts of moisture in the aggregate must be addressed in order to obtain accurate control.

1. Measuring cement. If bagged cement is used, the batches should be of such a size that only full bags are used. If this is not possible, then the proper amount should be weighed out each time. Volume is not an accurate method of measuring cement. Bulk cement (not bagged) should be weighed for each batch. Weigh a bag of Portland Type I cement and record the weight.   Completed ☐

2. Measuring water. The effects of the water-cement ratio on the qualities of concrete make it necessary to accurately measure the water. Water is measured in gallons. (One gallon of water weighs 8.33 lb.) Any method that will ensure accuracy is acceptable. Fill a five-gallon bucket, weigh it, and record the weight.   Completed ☐

3. Measuring aggregates. Measurement of aggregates by weight is the recommended practice. Aggregates should be measured by weight on all jobs that require a high degree of consistency in each batch.   Completed ☐

   Measurement of fine aggregate (sand) by volume is not accurate because a small amount of moisture is nearly always present. This moisture causes the fine aggregate to bulk or fluff up. The degree of bulking depends on the amount of moisture present and the fineness of the sand. A fine sand with a 5% moisture content will increase in volume about 40% above its dry volume.

   Weigh a cubic foot of very dry sand and a cubic foot of moist sand. Compare their weights. Use a measuring box.

4. Compare the results of measuring aggregates by volume and measuring by weight. Remember these comparisons.   Completed ☐

5. Clean up the area and return tools and materials to their assigned places.   Completed ☐

Instructor's Initials: _____

Date: _____

Job 47 Review

After completing this job successfully, answer the following questions.

1. Why is the practice of filling the concrete mixer with a certain number of shovelfuls of sand, cement, and coarse aggregate not recommended?

   _____

   _____

   _____

2. Volume is not an accurate method of measuring cement. Why?

   _____

   _____

   _____

3. How much does one gallon of water weigh?

   _____

4. About how much does fine sand increase in volume when it has a 5% moisture content?

   _____

5. How should the water to be used in a batch of concrete be measured?

   _____

   _____

   Score: _____

Name _____ Date _____ Class _____

JOB 48: Mixing Concrete with a Power Mixer

OBJECTIVE: After completing this job, you will be able to mix concrete in a power mixer using the proper technique.

TEXTBOOK REFERENCE: Study the appropriate section in Chapter 18, *Concrete Materials and Applications,* before starting this job.

Equipment

To complete this job, you will need the following tools and materials:

- Power cement mixer
- Shovel
- Measuring box or scales
- Five-gallon bucket
- Bag of Type I cement
- Source of dry sand
- Crushed stone
- Source of water
- Eye protection

Recommended Procedure

All concrete should be mixed thoroughly until it is uniform in appearance and all ingredients are uniformly distributed. The mixing time depends on several factors: the speed of the machine, the size of the batch, and the condition of the mixer.

Generally, the mixing time should be at least one minute for mixtures up to 1 cu yd with an increase of 15 seconds for each 1/2 cu yd or fraction thereof. Mixing time should be measured from the time all materials are in the mixer. Generally, about 10% of the mixing water is placed in the mixer before the aggregate and cement are added. Water should then be added uniformly along with the dry materials. The last 10% of the water is added after all the dry materials are added.

1. Assemble all of the materials and tools needed for this job. Arrange your workspace. Completed ☐

2. Check the mixer to be sure it is clean and in good working order. Start the mixer and add about 1/10 of the water anticipated for a 4 cu ft batch. (A typical batch proportion is 1:2:3; 1 part cement, 2 parts sand, and 3 parts crushed stone.) The amount of water needed for this mix is about 1/2 gal/cu ft of cement. Completed ☐

3. Add the sand and cement in proper amounts to the mixer. Add all but about 10% of the water. Mark the time. Completed ☐

4. Mix the ingredients for 15 seconds. Stop the mixer and examine the results. Decide whether or not to add any more water. Remember that concrete should only be as wet as is necessary to place and finish it. The more water that is added, the weaker the hardened product. Completed ☐

5. If more water is added, mix for another 5 seconds. Completed ☐

Go to Job 49 and perform a slump test on the plastic concrete.

Instructor's Initials: _____

Date: _____

Job 48 Review

After completing this job successfully, answer the following questions.

1. What is the intended use of Type I cement?
   _____
   _____
   _____

2. What factors is mixing time dependent on?
   _____
   _____
   _____

3. What is the recommended mixing time for a 1 cu yd batch?
   _____

4. In a typical batch, what is the proportion of cement to sand to large aggregate?
   _____

5. What is the effect on hardened concrete if too much water is used in the mixing process?
   _____
   _____

6. What is the normal curing time of concrete in days?
   _____

   Score: _____

Name _____ Date _____ Class _____

JOB 49: Performing a Slump Test on Plastic Concrete

OBJECTIVE: After completing this job, you will be able to perform a standard slump test on plastic concrete using the proper technique.

TEXTBOOK REFERENCE: Study the appropriate section in Chapter 18, *Concrete Materials and Applications,* before starting this job.

Equipment

To complete this job, you will need the following tools and materials:

- Eye protection
- Standard slump cone
- Metal rod 24″ long by 5/8″ in diameter
- Mason's trowel
- Five-gallon bucket with water
- Measuring tape or folding rule

Recommended Procedure

The mix consistency or degree of stiffness of plastic concrete is called *slump*. Slump is measured in inches. Very fluid (wet) mixes are called high-slump concrete, while stiff (dry) mixes are called low-slump concrete. Slump is related primarily to water-cement ratio. Generally, a low-slump concrete will produce a better concrete product. See **Figure 49-1**.

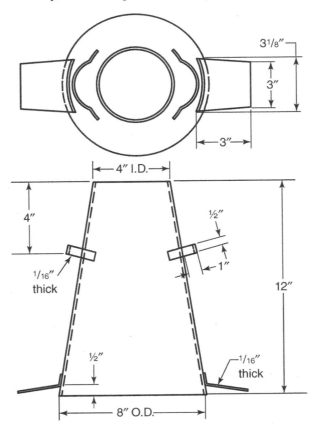

Goodheart-Willcox Publisher

Figure 49-1. Sheet metal slump cone.

1. Assemble the tools and materials needed for this job. Set up a work area close to the concrete mixer.   Completed ☐

2. Place the slump cone on a clean, dry, and level surface. Fill it 1/3 full of concrete from the freshly mixed batch. The sample should be representative of the total batch. Use a mason's trowel or shovel to take the sample.   Completed ☐

3. Using the 24″ metal rod, rod the concrete in the cone 25 times. Move the rod to different locations so the concrete is thoroughly mixed.   Completed ☐

4. Remove the rod and add more concrete until the cone is 2/3 full. Rod the concrete 25 more times, with the rod penetrating to the bottom.   Completed ☐

5. Fill the cone with concrete. Rake off the top with the rod until it is level. Then rod 25 more times as before.   Completed ☐

6. Remove the rod, remove the cone, and set it beside the concrete. Immediately measure the slump by placing the rod across the top of the cone and over the concrete. Measure the distance from the bottom of the rod to the top of the concrete. This distance is the slump rating for the batch.   Completed ☐

7. Clean up the tools and the area and return all tools and equipment to their assigned places. Be sure to remove all concrete from the mixer.   Completed ☐

Instructor's Initials: _____

Date: _____

Job 49 Review

After completing this job successfully, answer the following questions.

1. Define the term *slump*.

   _____
   _____

2. Describe the difference between concrete that has a high slump and concrete that has a low slump.

   _____
   _____
   _____

3. Why is concrete rodded so many times when filling a metal slump cone?

   _____
   _____
   _____

4. What are the dimensions of the rod that is used in a slump test?

   _____

5. How is the measurement made to determine the slump of a concrete batch?

   _____
   _____
   _____

Score: _____

Name _____ Date _____ Class _____

JOB 50: Placing Concrete in a Slab Form

OBJECTIVE: After completing this job, you will be able to place concrete in a slab form using the proper technique.

TEXTBOOK REFERENCE: Study the appropriate section in Chapter 20, *Concrete Flatwork and Formed Shapes,* before starting this job.

Equipment

To complete this job, you will need the following tools and materials:

- Prepared form ready for concrete
- Protective clothing
- Shovel
- Rake

Recommended Procedure

Concrete is moved about for placing by many methods, including chutes, push buggies, buckets handled by cranes, wheelbarrows, and pumped through pipes. The method used to move the concrete is not important as long as it does not impact the desired consistency of the concrete. Consistency should be governed by placing conditions and the application.

In this job, you will gain experience in placing concrete in a form designed to cast a flat slab. The size is not important—a section of sidewalk (3' × 6') would be adequate.

1. Assemble the tools and materials needed for this job. Wear protective clothing such as safety glasses, hard hat, gloves, and rubber boots if the slab is too large to work from the edges. Completed ☐

2. Check the form and subgrade to be sure everything is ready for the concrete. Once the concrete arrives, you will not have time to do these things. Forms should be in place and level. The subgrade should be smooth and moist. Moistening the subgrade prevents the rapid loss of water from the concrete when flatwork is being placed. This is especially important in hot weather. Refer to **Figure 50-1**. Completed ☐

Portland Cement Association

Figure 50-1. Form ready for concrete.

3. If reinforcing steel is to be used in the slab, it should be in place. Make sure the reinforcing steel is clean and free of loose rust or scale. No hardened mortar or concrete should be on the steel. Completed ☐

Copyright Goodheart-Willcox Co., Inc.
May not be reproduced or posted to a publicly accessible website.

4. When the concrete arrives, spread it in various parts of the form. It should not be placed in large quantities in one place and allowed to run, nor should it be worked over a long distance in the form. Segregation (separation) of the ingredients and sloping work planes result from these practices. Generally, concrete should be placed in horizontal layers having uniform thickness.

Completed ☐

Do not allow concrete to drop freely more than 3′ or 4′. Drop chutes of rubber or metal may be used when placing concrete in thin vertical sections.

5. Start placement of concrete in slab construction at the most distant point of the work so each batch can be dumped against the previously placed concrete, not away from it. Take care to prevent stone pockets (areas of excessive large aggregate) from occurring.

Completed ☐

6. Work the mix with a spade or rod to ensure that all spaces are filled and air pockets are worked out. This is called *puddling*, *spading*, or *rodding*. Mechanical vibrators can be used as well.

Completed ☐

7. Be sure there is enough concrete in the form for the finishing operation. The next procedure is finishing.

Completed ☐

Proceed to Job 51.

Instructor's Initials: _____

Date: _____

Job 50 Review

After completing this job successfully, answer the following questions.

1. What factors should govern the consistency of concrete?

   _____
   _____
   _____

2. What is the typical thickness of a sidewalk?

   _____

3. What is the desired condition of the subgrade for the placement of concrete for a flat slab?

   _____
   _____

4. What personal safety equipment should be used when working with concrete?

   _____
   _____

5. What does *segregation of concrete ingredients* mean?

   _____
   _____

6. What should you do immediately if you get cement in your eye?

   _____
   _____

Score: _____

Name _____ Date _____ Class _____

JOB 51 — Finishing Concrete Slabs

This job continues from Job 50.

OBJECTIVE: After completing this job, you will be able to finish a concrete slab using the proper technique.

TEXTBOOK REFERENCE: Study the appropriate section in Chapter 20, *Concrete Flatwork and Formed Shapes,* before starting this job.

Equipment
To complete this job, you will need the following tools and materials:
- Protective clothing
- Screed
- Darby or bull float
- Edgers
- Float
- Cement trowel
- Jointing tool
- Soft-bristled broom
- 1 × 10 board as wide as the slab

Recommended Procedure
Screeding is the process of striking off the excess concrete to bring the top surface to the proper grade or elevation. This is usually the first finishing operation after the concrete is placed in the forms. Screeding is performed with a screed. See **Figure 51-1**.

Sharon Meredith/iStock/Thinkstock

Figure 51-1. Excess concrete is struck off with a screed.

1. Screeding. Rest the screed on the top of the form and move it across the concrete in a sawing motion. Advance the screed forward slightly with each motion. This operation should be done before any bleeding (water rising to the surface) occurs. Screed the surface. Completed ☐

2. Darbying or bull floating. Floating is performed with a darby or bull float to eliminate high and low spots and embed large aggregate. See **Figure 51-2**. This operation follows immediately after screeding to prevent bleeding. Use the darby or bull float on the surface. Completed ☐

Marshalltown Company

Figure 51-2. A bull float is used to remove high and low spots.

3. Edging. Edging is frequently the next operation. Edging provides a rounded edge or radius to prevent chipping or damage to the edge. Run the edger back and forth until the desired finish is obtained. See **Figure 51-3**. Edge the slab. Completed ☐

Stanley Goldblatt

Figure 51-3. Edging produces a radius on the edge of the slab.

Name _____

4. Jointing. Jointing is performed after edging. The cutting edge (bit) of the jointer cuts a groove in the slab to form a control or contraction joint. Cracking will occur at this point. Tooled joints are usually placed at intervals equal to the width of the slab, but not more than 10′ apart in sidewalks and driveways. Use a straightedge as a guide when making a groove. See **Figure 51-4**. Joint the surface.

Completed ☐

Stanley Goldblatt

Figure 51-4. Jointing a slab.

5. Floating. After concrete has been edged and jointed, it should be allowed to harden enough to support a person and leave only a slight foot imprint. Floating should not begin until the water sheen has disappeared. The surface is floated with wood or metal floats or with a finishing machine using float blades. A light metal float forms a smoother surface texture than a wood float. Float the surface.

Completed ☐

6. Steel troweling. When a smooth, dense surface is desired, steel troweling is performed after floating. Masons frequently float and trowel an area before moving their kneeboards. As the concrete hardens, it may be troweled several times to obtain a very smooth and hard surface. See **Figure 51-5**.

Completed ☐

John Sarlin/iStock/Thinkstock

Figure 51-5. Finish troweling a concrete slab. This worker is using two trowels in order to put extra weight on the trowel and for added balance.

If necessary, tooled joints and edges may be rerun after troweling to maintain uniformity. Some slabs are safer when broomed or brushed to produce a slightly roughened surface. This can be done with a soft-bristled push broom after the steel troweling. Brooming is usually done perpendicular to the traffic direction

Instructor's Initials: _____

Date: _____

Job 51 Review

After completing this job successfully, answer the following questions.

1. What is the purpose of screeding?

   _____
   _____
   _____

2. What tools are used to perform the second finishing operation on a concrete slab?

   _____
   _____

3. What is the purpose of edging a concrete slab?

   _____
   _____
   _____

4. Why is a concrete slab jointed?

   _____
   _____
   _____

5. When should floating begin?

   _____
   _____
   _____

6. What is the last operation in finishing a concrete slab when a smooth, dense surface is desired?

   _____

Score: _____

Name _____ Date _____ Class _____

JOB 52 — Building Footing Forms for Concrete

OBJECTIVE: After completing this job, you will be able to build footing forms for concrete using the proper technique.

TEXTBOOK REFERENCE: Study the appropriate section in Chapter 19, *Form Construction,* before starting this job.

Equipment

To complete this job, you will need the following tools and materials:

- Mason's or carpenter's hammer
- Saw
- Supply of 16-penny nails
- Seven 2″ × 4″ × 12″ stakes
- Three 2″ × 4″ × 8′ boards
- Level
- Framing square
- Flexible tape or folding rule
- Pencil
- Shovel
- Small sledgehammer

Recommended Procedure

Wood is the most popular form material. Both lumber and plywood are generally used in form construction. Forms must be strong enough to resist the forces developed by the plastic (liquid) concrete. Regular concrete weighs about 150 lb/cu ft. Safety is always a consideration when building forms.

Forms should be designed so they are practical, economical, and in the correct shape, width, and height. They must be able to retain their shape during the placing and curing phases. Wet concrete should not leak from joints and cause fins and ridges. Forms must also be designed so they can be removed without damaging the concrete.

Footings are not usually visible, so their appearance is of little concern. However, footings must be located accurately and built to the specified dimensions. Concrete footing forms are usually constructed from 2″ construction lumber the same width as the footing thickness. The boards are placed on edge as shown in **Figure 52-1** and held in place by stakes and cross-spreaders.

1. This job requires you to build an *L*-shaped footing form that is 16″ wide. Each leg of the footing, which is measured along the outside of the *L*, is 48″ long. The material is standard 2″ × 4″ construction lumber. Stakes are 2″ × 4″ pieces sharpened on the end. The stakes should be 12″ long. Prepare the materials.

   Completed ☐

2. Using **Figure 52-1** as the pattern, cut your materials to length and prepare the stakes.

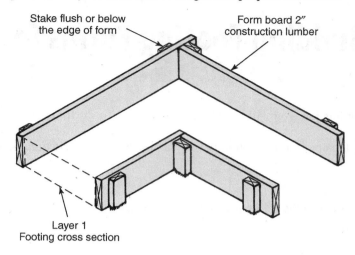

Figure 52-1. Footing form made from 2″ construction lumber.

Goodheart-Willcox Publisher

3. Check with your instructor to select the location for your footing form. Lay out the outside form boards and drive two stakes into the ground about 9″ deep. Place the outside form board against the stake and hold it with your foot while you drive a nail through the stake and then into the board deep enough to hold it in place. The top of the stake should be slightly below the top edge of the form board.   Completed ☐

4. Go to the other stake, place the level on the form board, and mark the point on the stake where the form board is level. Remove the level and nail the stake to the form board. Use two nails. Go to the first stake and complete the process there. Use the shovel to level the ground if there is a high spot.   Completed ☐

5. Check the board to see if it is still level. If not, drive the stake down at the high end.   Completed ☐

6. Proceed to build the rest of the form using this approach. Check regularly to see that the form is level and the proper width and length.   Completed ☐

7. When all the boards are in place, check all dimensions and be sure the form is level across as well as along the form.   Completed ☐

8. When you are satisfied the form is correct, ask your instructor to inspect it.   Completed ☐

9. Take the form apart, remove all nails, and return all tools and materials to their proper places.   Completed ☐

Instructor's Initials: _____

Date: _____

Job 52 Review

After completing this job successfully, answer the following questions.

1. How much does regular concrete weigh?

_____

2. In addition to retaining their shape and not leaking, what is another major consideration in the design and construction of concrete forms?

_____

_____

Name _____

3. What is the most popular material used to make site-made concrete forms?
   _____

4. How can curved shapes be formed using wood forms?
   _____
   _____
   _____
   _____

5. Why should the stakes *not* project above the top edge of concrete slab forms?
   _____
   _____

Score: _____

Notes

Name _____ Date _____ Class _____

JOB 53: Building Wall Forms for Concrete

OBJECTIVE: After completing this job, you will be able to build a wall form for concrete using the proper technique.

TEXTBOOK REFERENCE: Study the appropriate section of Chapter 19, *Form Construction*, before starting this job.

Equipment

To complete this job, you will need the following tools and materials:

- Mason's or carpenter's hammer
- Supply of 6- and 16-penny nails
- Flexible tape or folding rule
- Supply of 2″ × 4″ lumber
- Supply of 3/4″ exterior plywood
- Four crimp snap-in form ties
- Concrete nails or power-driven nails
- Saw
- Level
- Framing square
- Pencil
- Power drill
- Safety equipment

Recommended Procedure

In this job, you will build a wall form 32″ high by 48″ long for a concrete wall 10″ thick. The wall form should be anchored to a footing (or floor) as shown in **Figure 53-1**.

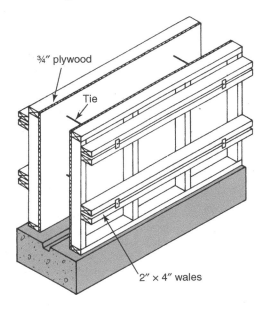

Goodheart-Willcox Publisher

Figure 53-1. Typical site-made wall form.

1. To build an in-place wall form, first cut two sole plates, two top plates, and eight studs. The plates should be 48″ long. The studs should be 29″ long. (A 2″ × 4″ stud is actually 1 1/2″ × 3 1/2″.) Completed ☐

2. Lay out the pieces and nail them together with 16-penny common nails. Use two nails per connection. Be sure the studs are spaced 16″ apart and the frame is square. Build two identical frames. Completed ☐

3. Cut two pieces of sheathing to fit the frames (32″ × 48″). Nail the sheathing to the frames using 6-penny box nails. Completed ☐

4. Mark the location on the footing where the forms will be located to produce a 10″-thick concrete wall. The form should be centered on the footing. Completed ☐

WARNING!
For Step 5, be sure to wear safety glasses, a hard hat, and other appropriate safety equipment. Also, do not use a power-activated nailer until you have had instruction in its proper use.

5. Ask someone to help you lift each frame to its location on the footing. Attach each frame to the footing using concrete nails or power-driven nails. Completed ☐

6. Determine the four locations of the form ties. Refer to **Figure 53-1**. Drill a hole at each location large enough to receive the tie. Repeat the operation on the other form. Completed ☐

7. Insert the form ties through the holes according to the manufacturer's instructions. Completed ☐

8. Cut eight 2″ × 4″ × 48″ wales. Ask someone to help you hold a pair of wales in place while you tighten the snap tie. Wales are not nailed to the form. Complete the process by attaching all of the ties to the wales. Completed ☐

9. Check the assembly for proper spacing and then be sure it is level and plumb. Make any necessary adjustments. Completed ☐

10. Clean up the work area and return all tools and materials to their assigned places. Completed ☐

Save this assembly for Job 54.

Instructor's Initials: _____

Date: _____

Job 53 Review

After completing this job successfully, answer the following questions.

1. Name the five basic structural parts of a wall form.

   _____

   _____

2. How is the proper spacing maintained between the two sides of a typical site-constructed wall form?

   _____

   _____

3. How are the wales attached to the wall forms?

   _____

   _____

Name _____

4. What methods are generally used to attach the sole plate to the footing?
   _____
   _____

5. What kind of plywood is used for sheathing in a wood wall form?
   _____
   _____

Score: _____

Notes

Name _____ Date _____ Class _____

JOB 54: Building and Installing a Buck

OBJECTIVE: After completing this job, you will be able to build and install a buck in a wall form using the proper technique.

TEXTBOOK REFERENCE: Study the appropriate section of Chapter 19, *Form Construction*, before starting this job.

Equipment

To complete this job, you will need the following tools and materials:

- Mason's hammer or carpenter's hammer
- Saw
- Supply of 6-penny nails
- Level
- Framing square
- Flexible tape or folding rule
- Pencil
- Supply of 1″ × 12″ pine boards or 3/4″ plywood
- Safety equipment

Recommended Procedure

In this job, you will build a 12″ × 18″ buck to be placed in the wall form constructed in Job 53. Bucks are wood or steel frames set in the form between the inner and outer form to make an opening in the wall. Any openings in the wall, such as windows, can be formed using bucks. See **Figure 54-1**.

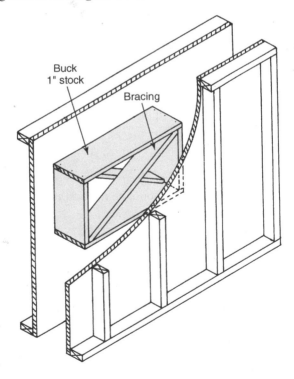

Goodheart-Willcox Publisher

Figure 54-1. A buck in a wall form.

1. To build a buck for placement in a wall form, first lay out the lengths needed and check your dimensions before cutting the pieces. The buck will be 12″ high by 18″ wide (outside dimensions). Do not forget to subtract the material thickness for the end pieces. Completed ☐

2. Cut the four sides of the buck to the proper length and width. Check the distance between the forms to be sure your buck will fit snugly between them. Completed ☐

3. Nail the frame together, making sure that the corners are perpendicular (90°). Check the assembly with the framing square. Completed ☐

4. Measure the inside diagonal distance from corner to corner and cut two 6″-wide boards to serve as braces. Study **Figure 54-1**. Cut the pieces when you are sure the dimensions are correct. Completed ☐

5. Nail the braces in place. Be careful not to rack (distort) the buck during this operation. Completed ☐

6. Place the buck in the wall form and secure it with four nails on each side of the buck. Do not drive these nails all the way so they can be pulled before removing the wall forms. (Special form nails are made for this purpose.) Make sure the buck is level in the form. Locate the nails by measuring from the end and top of the form. Completed ☐

7. Inspect your work, clean up the area, and return all tools and materials to their assigned places. Completed ☐

Instructor's Initials: _____

Date: _____

Job 54 Review

After completing this job successfully, answer the following questions.

1. What are bucks and what purpose do they serve?

   _____
   _____
   _____

2. What technique is used in wood buck construction to prevent racking (distortion)?

   _____
   _____

3. How can you recognize a form nail?

   _____

4. Based on the experience gained from this job, when would be the best time to install a buck between the forms?

   _____
   _____
   _____

5. When a buck is being built, what dimensions are most critical?

   _____
   _____

Score: _____

Name _____ Date _____ Class _____

JOB 55 / Installing Round Column Forms

OBJECTIVE: After completing this job, you will be able to install a round column form using the proper technique.

TEXTBOOK REFERENCE: Study the appropriate section in Chapter 19, *Form Construction*, before starting this job.

Equipment

To complete this job, you will need the following tools and materials:

- Mason's or carpenter's hammer
- Saw
- Supply of 16-penny nails
- Level
- Flexible tape or folding rule
- Pencil
- Supply of 2″ × 4″ material
- Safety equipment
- Round paper column form

Recommended Procedure

Round column forms are usually metal, fiberboard, or paper. See **Figure 55-1**. Round prefabricated forms are produced for various lengths and diameters. Forms made from fiberboard and paper are not reusable but have some advantages. They are economical and lightweight, and the paper surfaces produce a very smooth concrete surface with no seams. In this job, you will install a paper form to produce a round concrete column.

 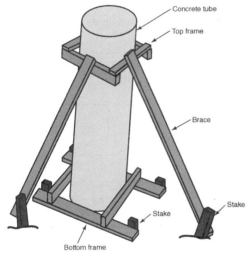

Goodheart-Willcox Publisher

Figure 55-1. Left—A paper column form. Right—A column form with scaffold support.

1. Identify the location where the column form is to be positioned. The surface should be a footing or over a hole in the ground. Completed ☐

2. Examine the column form to be sure it is not damaged and is the proper diameter and length for the job. If the form is too long, it can be cut with a hand saw or power saw. Completed ☐

3. Using 2×4 lumber and double-headed nails, build a scaffold type support system around the column form as shown in **Figure 55-1**. The framework should be a sturdy structure, since it will serve as the bracing for the column. Remember, if this project was real, the tube form filled with concrete would weigh several hundred pounds. Do not nail into the form itself. Completed ☐

Copyright Goodheart-Willcox Co., Inc.
May not be reproduced or posted to a publicly accessible website.

4. Stake the bottom frame in-place so the tube cannot move in either direction. Use a 4′ level to determine plumb of the form tube. Place the level on the tube, directly underneath one of the upper braces. **Make sure that the level is in a vertical plane with the form tube.** Use the level and move the tube until it reads plumb. Then tack the brace to temporarily secure the tube from moving and repeat the process with the next brace. Continue to make minor adjustments until both locations under the braces indicate that the level readings are plumb. Then nail the braces permanently to secure the structure.   Completed ☐

5. Inspect the total assembly to be sure it is safe and well-constructed.   Completed ☐

6. Clean up the area and return any unused materials, tools, and equipment to their proper places.   Completed ☐

Instructor's Initials: _____

Date: _____

Job 55 Review

After completing this job successfully, answer the following questions.

1. What materials are frequently used for round concrete column forms?
   _____
   _____

2. What type of round concrete column forms are reusable?
   _____

3. Identify three advantages of forms made from fiberboard and paper.
   _____
   _____
   _____

4. What type of material is generally used to brace a paper concrete column form?
   _____
   _____

5. How can paper forms be cut to length?
   _____

Score: _____

Name _____ Date _____ Class _____

JOB 56: Building Centering for a Masonry Arch

OBJECTIVE: After completing this job, you will be able to build centering for a masonry arch using the proper technique.

TEXTBOOK REFERENCE: Study the appropriate section in Chapter 16, *Wall Systems,* before starting this job.

Equipment

To complete this job, you will need the following tools and materials:

- Mason's or carpenter's hammer
- Saws
- Flexible tape or folding rule
- Pencil
- Supply of 1″ × 2″ pine boards
- Two 2″ × 8″ × 4′ boards
- Supply of 8-penny box nails
- Cardboard arch pattern
- Square
- Mason's line

Recommended Procedure

Arch centering is designed to support the masonry while it is being constructed. Arch centering is usually made from wood. The ribs are 2″ construction lumber, and the lagging may be 1″ × 2″ pine strips. The lagging should be cut 1″ shorter than the thickness of the masonry wall so it does not interfere with the mason's line. The ribs are cut to the shape of the arch. The centering must be sturdy to support the weight of the masonry units. See **Figure 56-1**.

In this job, you will construct the centering for a segmental arch in an 8″-thick brick masonry wall. The width of the arch is 36″.

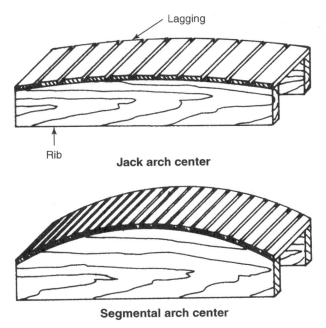

Goodheart-Willcox Publisher

Figure 56-1. Centering for an arch.

Copyright Goodheart-Willcox Co., Inc.
May not be reproduced or posted to a publicly accessible website.

1. If a cardboard pattern of the arch is available, examine it to be sure it is the proper design and size. If a pattern is not available, then make one. The total width should be 36″. The rise may be 1/6, 1/8, 1/10, or 1/12 of the span. We will use 1/12 for this job. Therefore, the rise will be 3″. See **Figure 56-2**. Make a pattern using the mason's line and a pencil.   Completed ☐

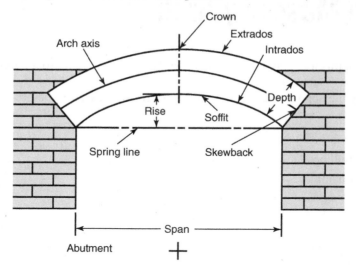

Figure 56-2. Arch terminology.

Goodheart-Willcox Publisher

2. Using your pattern, lay out the arch intrados (lower curve of the arch) on one of the 2″ × 8″ rib boards after they have been cut to a length of 36″. Position the pattern so the top of the curve is almost to the edge of the board. Mark out the curve on the rib board.   Completed ☐

3. Saw out the curve using a jigsaw or band saw. Be sure you are familiar with these tools before using them.   Completed ☐

4. Using the same process, lay out the curve on the other rib board. These two boards must be identical. Cut out this curve.   Completed ☐

5. Using the 1″ × 2″ material, cut several strips 7″ long to form the lagging. Two of them should be beveled on the edge to continue the vertical end of the centering.   Completed ☐

6. Nail several lagging strips into place, leaving 1/2″ to 3/4″ between them. Cut the remaining strips and nail them to the rib boards.   Completed ☐

7. Check over your work, clean up the area, and return all tools and materials to their assigned places.   Completed ☐

Instructor's Initials: _____

Date: _____

Job 56 Review

After completing this job successfully, answer the following questions.

1. What is the purpose of arch centering?

_____

_____

Name _____

2. What is *lagging*?
   _____
   _____

3. How is the shape of the ribs determined?
   _____
   _____

4. What would the rise of a segmental arch be if it were 48″ wide and had a rise of 1/8 the span?
   _____

5. Why was a jigsaw or band saw recommended to cut the curve on the centering ribs?
   _____
   _____

6. Why are narrow strips used for lagging?
   _____
   _____

Score: _____

Notes

Name _____ Date _____ Class _____

JOB 57: Dampproofing Concrete Block Basement Walls

OBJECTIVE: After completing this job, you will be able to dampproof a concrete block basement wall using the proper technique.

TEXTBOOK REFERENCE: Study the appropriate section in Chapter 15, *Foundation Systems,* before starting this job.

Equipment

To complete this job, you will need the following tools and materials:
- Plasterer's trowel
- Supply of Portland cement plaster or mortar
- Wire scratcher or notched trowel
- Bituminous dampproofing compound or cement-based paint
- Mason's trowel
- Stiff-bristle brush
- Hard hat, safety glasses, and other appropriate safety equipment

Recommended Procedure

1. Apply a 1/4"-thick parge coat of Portland cement plaster or mortar to the outside of a clean and damp concrete block basement wall. Use a plasterer's rectangular trowel for this operation. The coat should be worked into the cracks and crevices to form a watertight barrier. See **Figure 57-1**.

Author's image taken at Job Corps, Denison, IA

Figure 57-1. Applying the first parge coat to a basement wall.

2. Allow the first parge coat to partially harden and then roughen it with a wire scratcher or notched trowel to provide a good bond for the second coat. See **Figure 57-2**. Keep the first coat moist for 24 hours.

Completed ☐

Author's image taken at Job Corps, Denison, IA

Figure 57-2. Using the notched trowel to roughen the first parge coat.

3. After 24 hours, the first parge coat should be sufficiently cured to apply the second coat. Apply another 1/4″-thick parge coat to the wall. Apply firm pressure and leave the surface smooth and dense. See **Figure 57-3**.

Completed ☐

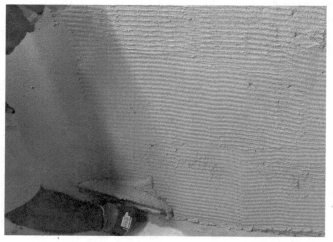

Author's image taken at Job Corps, Denison, IA

Figure 57-3. Applying the second parge coat.

4. Form a cove over the footing to prevent water from collecting at this point. The second coat should be allowed to cure for at least 48 hours before applying any further dampproofing.

Completed ☐

5. For added moisture resistance, you can apply a heavy coat of tar, two coats of a cement-based paint, or a covering of plastic film to the wall. Follow manufacturer's instructions when using any commercial product.

Completed ☐

Instructor's Initials: _____

Date: _____

Name _____

Job 57 Review

After completing this job successfully, answer the following questions.

1. What material is commonly used as a parge coat to dampproof a concrete block basement wall?
   _____

2. How many coats of parging are recommended to dampproof a concrete block basement wall?
   _____

3. What tool is recommended to apply a parge coat?
   _____
   _____

4. Why is a scratcher or notched trowel used on the first parge coat?
   _____
   _____

5. What is the function of the cove between the wall and the footing?
   _____
   _____

6. What types of materials may be used over a parged wall to increase resistance to moisture penetration?
   _____
   _____
   _____

   Score: _____

Notes

Name _____ Date _____ Class _____

JOB 58: Building Columns, Piers, and Pilasters

OBJECTIVE: After completing this job, you will be able to build brick and block columns, piers, and pilasters using proper technique.

TEXTBOOK REFERENCE: Study the appropriate section in Chapter 15, *Foundation Systems,* before starting this job.

Equipment

To complete this job, you will need the following tools and materials:

- Mason's tools
- Mortar
- Supply of bricks and blocks
- Hard hat, safety glasses, and other appropriate safety equipment

Recommended Procedure

Columns, piers, and pilasters transmit loads to the footing. Columns and piers are freestanding, but pilasters are built into the foundation wall. This job will provide an opportunity to build a column, a pier, and a pilaster using brick and block.

Constructing a 12″ × 12″ Brick Column

1. Assemble the materials needed to build a 12″ × 12″ brick column on a footing to a height of 12 courses. (A column's height is at least three times its thickness.) See **Figure 58-1**.   Completed

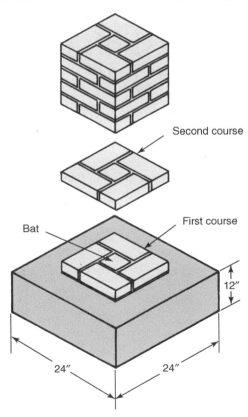

Goodheart-Willcox Publisher

Figure 58-1. Bonding for a 12″ × 12″ brick column.

2. Lay the first course of bricks on the footing in the designated location. Use full bed joints and check to see that this course is level, plumb, and square. Completed ☐

3. Using the bond pattern illustrated for the second course, lay the second course. Remove excess mortar and level this course. Completed ☐

4. Complete the column to 12 courses. The structure should be a solid mass of bricks and mortar that is square, plumb, and level. Tool the joints and brush the bricks. Completed ☐

Constructing a 16″ × 16″ Concrete Block Pier

1. Assemble the materials needed to construct a 16″ × 16″ concrete block pier to a height of three courses. Completed ☐

2. Mark the location for the pier. Lay down a solid mortar bed and place the first course. See **Figure 58-2**. Completed ☐

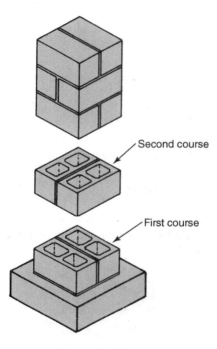

Goodheart-Willcox Publisher

Figure 58-2. Concrete block pier, 16″ × 16″ square.

3. Lay the second and third courses, alternating the position of the blocks as shown in the figure. The third course should be solid top blocks. Completed ☐

4. Tool the joints and brush the blocks. Completed ☐

Name _____

Constructing a 4″ × 12″ Brick and Block Pilaster

1. Assemble the materials needed to build an 8″ × 44″-long concrete block wall that incorporates a 4″ × 12″ brick pilaster. The wall height will be 32″ high. See **Figure 58-3**. Completed ☐

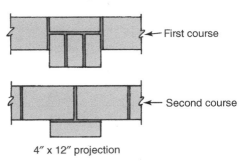

Goodheart-Willcox Publisher

Figure 58-3. Brick pilaster in a concrete block wall.

2. Snap a line to fix the location of the wall. Lay down a full bed of mortar and lay a regular 8″ × 8″ × 16″ concrete block at either end of the wall. The remaining space should accommodate a 12″-long brick. Completed ☐

3. Fill in the first course of bricks for the pilaster as shown in **Figure 58-3**. Check the first course of blocks and bricks to be sure they are level, plumb, and straight. Completed ☐

4. Lay two more courses of bricks to bring the pilaster up to the height of the course of blocks. Alternate the course pattern. Completed ☐

5. Lay the next course of blocks and fill in the bricks as you proceed. Study the figure of the second course. You will need to cut some blocks to make the pattern. Completed ☐

6. Lay the third course of blocks and bricks in the same manner as the first course. Tool the joints and brush the wall. Completed ☐

7. Clean up the area and return the tools and materials to their assigned places. Completed ☐

Instructor's Initials: _____

Date: _____

Job 58 Review

After completing this job successfully, answer the following questions.

1. What is the function of columns, piers, and pilasters?

   _____

   _____

2. Which is generally taller, a column or a pier?

   _____

3. Why were solid top blocks used as the final course on the concrete block pier that you constructed?

   _____

   _____

4. Why were bricks used to build the 4″ × 12″ pilaster in the concrete block wall?

_____
_____
_____

5. What type of bonding was used in the pilaster to increase its strength?

_____
_____

Score: _____

Name _____ Date _____ Class _____

JOB 59: Building Solid Masonry Walls

OBJECTIVE: After completing this job, you will be able to build solid masonry walls of brick and block using the proper technique.

TEXTBOOK REFERENCE: Study the appropriate section in Chapter 16, *Wall Systems,* before starting this job.

Equipment

To complete this job, you will need the following tools and materials:
- Mason's tools
- Mortar
- Supply of bricks and blocks
- Hard hat, safety glasses, and other appropriate safety equipment

Recommended Procedure

There are six generally accepted types of masonry walls—solid walls, 4″ RBM curtain and panel walls, hollow walls, anchored veneer walls, composite walls, and reinforced walls. This job involves solid masonry walls.

A solid masonry wall is built of masonry units laid close together with all joints between them filled with mortar. Solid masonry units, hollow masonry units, or a combination of the two may be used.

This job will provide an opportunity to construct two types of solid masonry walls—one using solid units (brick) and one using hollow units (block).

Solid Brick Masonry Wall with Headers

1. Assemble the materials necessary to build the solid brick masonry demonstration wall shown in **Figure 59-1**.　　　　　　　　　　　　　　　　　　　　　　　　　　　　　　　　Completed ☐

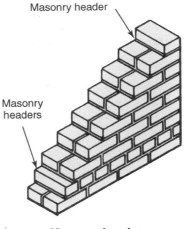

Goodheart-Willcox Publisher

Figure 59-1. A solid brick masonry wall with masonry headers.

2. Lay both wythes of the first course of bricks in running bond. Use a full mortar bed and fill the collar joint between the wythes.　　　　　　　　　　　　　　　　　　　　Completed ☐

3. Lay the second course as headers. Refer to **Figure 59-1**.　　　　　　　　　Completed ☐

4. Lay the next six courses in running bond identical to the first course. Keep your work neat and clean. Completed ☐

5. Lay the final course as headers similar to the second course. Tool the joints and brush the wall. Completed ☐

Solid Masonry Wall Using Hollow Units Masonry Bonded

1. Assemble the materials needed to build the 12″, thick demonstration wall shown in **Figure 59-2**. Completed ☐

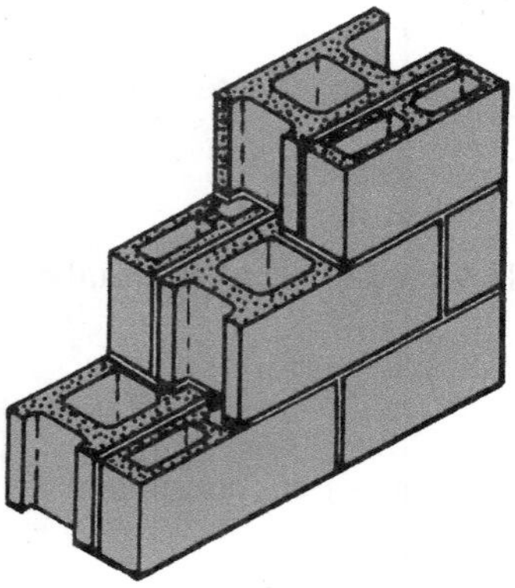

Masonry bonded

Goodheart-Willcox Publisher

Figure 59-2. A solid masonry wall using hollow concrete block.

2. Lay both wythes of blocks (4″ and 8″ blocks) in running bond on a solid mortar bed. Fill the collar joint between the wythes. Completed ☐

3. Lay the back wythe of the second course. Parge the face to form the collar joint between the wythes. Be sure to alternate the 4″ and 8″ blocks to develop the masonry bond. Completed ☐

4. Lay the front wythe and add more mortar to the collar joint, if needed. Completed ☐

5. Complete the demonstration wall by laying the third course identical to the first. Tool the joints and brush the wall. Completed ☐

6. Inspect your work, clean up the area, and return all tools and materials to their assigned places. Completed ☐

Instructor's Initials: _____

Date: _____

Job 59 Review

After completing this job successfully, answer the following questions.

1. List the six types of masonry walls.

_____

_____

_____

_____

Name _____

2. Is it permissible to use hollow masonry units to build a solid masonry wall? Why?

   _____
   _____

3. What is the function of the collar joint in a solid masonry wall?

   _____
   _____

4. What size concrete block units are used to build a 12″-thick solid masonry wall?

   _____

5. Which wythe was laid first in the 12″-thick solid masonry wall?

   _____

   Score: _____

Notes

Name _____ Date _____ Class _____

JOB 60: Building a 4″ RBM Curtain or Panel Wall

OBJECTIVE: After completing this job, you will be able to build a 4″ RBM (reinforced brick masonry) panel or curtain wall using proper technique.

TEXTBOOK REFERENCE: Study the appropriate section of Chapter 16, *Wall Systems,* before starting this job.

Equipment

To complete this job, you will need the following tools and materials:

- Mason's tools
- Mortar
- Supply of bricks
- Steel reinforcing
- Hard hat, safety glasses, and other appropriate safety equipment
- Flashing and wicking

Recommended Procedure

A curtain wall is an exterior nonloadbearing wall not wholly supported at each story. A panel wall is an exterior nonloadbearing wall supported at each story. Both walls must be able to resist lateral forces, such as wind pressures, and transfer these forces to adjacent structural members. **Figure 60-1** shows a typical 4″ RBM wall.

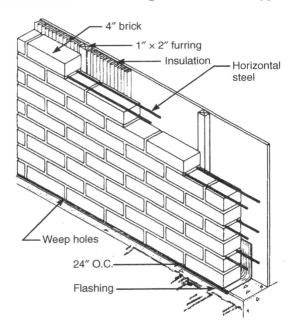

Brick Industry Association

Figure 60-1. Typical 4″ RBM curtain or panel wall.

Ladder-type or truss-type joint reinforcement is used throughout the length of the wall. Reinforcement must be completely embedded in mortar, not laid directly on top of the brick. All head and bed joints must be full to resist wind-driven rain.

1. Assemble the materials needed for this job. Assume the demonstration wall will be 6′ long, 4″ thick, and nine courses high. Lay the bricks to No. 6 on the modular rule. Completed ☐

2. Position the flashing along a chalk line or edge of a slab or foundation brick ledge. Lay down a full mortar bed for the first course, but allow for weep holes every 24″ along the wall. Lay the first course of stretchers. Completed ☐

3. Lay the second course in running bond on a full bed of mortar. Position the joint reinforcement (ladder-type or truss-type) on small pads of mortar so it will be in the center of the mortar joint. Completed ☐

4. Lay the third course of bricks on a full bed of mortar, being careful not to disturb the reinforcement. Completed ☐

5. Repeat the process, inserting reinforcing in every other bed joint. See **Figure 60-1**. Continue until the wall is nine courses high. Completed ☐

6. Tool the joints and brush the wall. Examine your work to be sure it is high quality. Completed ☐

7. Clean up the area and return the tools and materials to their assigned places. Completed ☐

Instructor's Initials: _____

Date: _____

Job 60 Review

After completing this job successfully, answer the following questions.

1. What does *RBM* represent?

   _____

2. What is the difference between a curtain wall and a panel wall?

   _____
   _____
   _____

3. Why is reinforcing required in single-wythe curtain and panel walls?

   _____
   _____

4. What type of reinforcing was used in your RBM wall?

   _____
   _____

5. Is it possible to lay a single-wythe brick wall in common bond?

   _____
   _____

6. What determines how frequently joint reinforcing should be placed in an RBM wall?

   _____
   _____

Score: _____

Name _____ Date _____ Class _____

JOB 61: Building a Hollow Brick Masonry Bonded Wall

OBJECTIVE: After completing this job, you will be able to build hollow masonry bonded walls of brick and block using the proper technique.

TEXTBOOK REFERENCE: Study the appropriate section of Chapter 16, *Wall Systems,* before starting this job.

Equipment

To complete this job, you will need the following tools and materials:

- Mason's tools
- Chalk line
- Mortar
- Supply of bricks and blocks
- Hard hat, safety glasses, and other appropriate safety equipment

Recommended Procedure

Hollow masonry walls are built using solid or hollow masonry units. The units are separated to form an inner and an outer wall. These walls may be either a cavity or masonry-bonded type wall. This job involves masonry-bonded walls. You will construct a hollow masonry wall using solid units (bricks).

This demonstration wall will be 8″ thick and will use the pattern shown in **Figure 61-1**. Notice the stretcher courses are laid as rowlock stretchers, while the alternate courses are laid as rowlock headers.

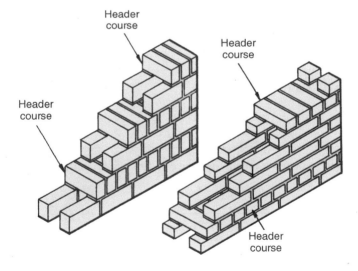

Goodheart-Willcox Publisher

Figure 61-1. Hollow brick masonry bonded walls.

1. Assemble the materials needed for this job. This wall will be 8″ thick, 32″ long, and six courses high. Completed ☐

2. Snap two chalk lines where the two sides of the wall will be located. Lay a bed of mortar for the inside wythe. Completed ☐

Copyright Goodheart-Willcox Co., Inc.
May not be reproduced or posted to a publicly accessible website.

3. Lay the inner wythe of brick as rowlock stretchers. Remove excess mortar between the wythes. Lay the outer wythe to the line. Check the distance between the faces of the wall to be sure it equals the length of a brick.   Completed ☐
4. Lay the second course as rowlock headers. Use full head joints.   Completed ☐
5. Repeat the pattern until the wall is six courses high.   Completed ☐
6. Strike the joints and brush the wall.   Completed ☐
7. Inspect your work, clean up the area, and return all tools and materials to their assigned places.   Completed ☐

Instructor's Initials: _____

Date: _____

Job 61 Review

After completing this job successfully, answer the following questions.

1. What wall type is similar to a hollow masonry bonded wall?
   _____

2. What is the basic difference between a hollow masonry bonded wall and a cavity wall?
   _____
   _____
   _____
   _____

3. Which wythe was laid first on the hollow masonry bonded wall?
   _____
   _____

4. How was the width of the wall determined?
   _____
   _____

5. Rowlock stretchers were used to provide the cavity space in this wall. Could regular headers have been used? Why?
   _____
   _____
   _____
   _____

Score: _____

Name _____ Date _____ Class _____

JOB 62: Building an Anchored Veneered Wall and Installing Flashing

OBJECTIVE: After completing this job, you will be able to build an anchored veneered wall and install flashing using the proper technique.

TEXTBOOK REFERENCE: Study the appropriate section of Chapter 16, *Wall Systems,* before starting this job.

Equipment

To complete this job, you will need the following tools and materials:

- Mason's tools
- Mortar
- Supply of bricks
- Supply of corrugated fasteners
- Flashing
- Hard hat, safety glasses, and other appropriate safety equipment
- Section of frame wall

Recommended Procedure

Facing veneer is attached to the backing but does not act structurally with the rest of the wall. Anchored brick veneer construction consists of a nominal 3″- or 4″-thick exterior brick wythe anchored to the backing system (frequently a frame wall). The wythe is anchored with metal ties in such a way that a clear air space is provided between the veneer and the backing system. The backing system may be wood frame, steel frame, concrete, or masonry.

Brick or stone veneer on a frame backing transfers the weight of the veneer to the foundation. The foundation brick ledge should be at least equal to the total thickness of the brick veneer wall assembly.

There should be a tie for every 2 2/3 sq ft of wall area, with a maximum spacing of 24″ O.C. in either direction. The best location for the nail is at the bend (not greater than 5/8″ away from the bend) in the corrugated tie, and the bend should be 90°. Corrugated ties must also penetrate to at least half the veneer thickness and be completely embedded in the mortar.

In this job, you will build a section of brick or stone veneered wall with a typical frame wall backup. The length of the section is 48″ and the height is nine courses.

1. Assemble the tools and materials needed for the job. Arrange a convenient work area. See **Figure 62-1**.   Completed ☐

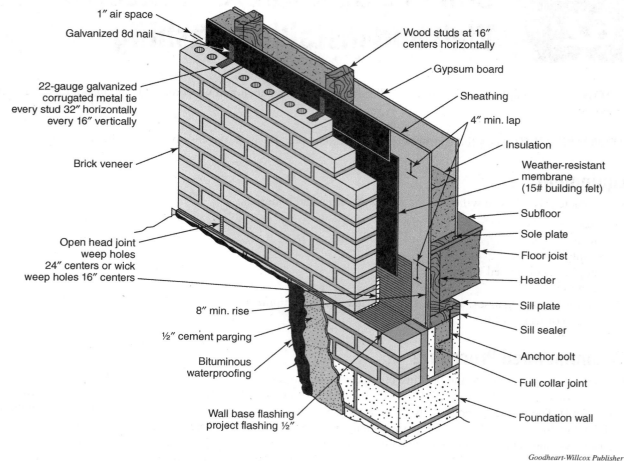

Figure 62-1. Typical brick veneer wall with wood frame backup.

2. Snap a chalk line to locate the face of the veneer if the demonstration wall is on the floor. Place the flashing in its proper location and lay a full bed of mortar.   Completed ☐

3. Lay the first course of brick or stone maintaining a 1″ air space between the veneer and the backup. Install two weep holes about 24″ apart by omitting the head joints.   Completed ☐

4. Straightedge and level the course. Lay the second course of masonry units on a full mortar bed. Be sure the weep holes remain open.   Completed ☐

Name _____

5. Attach a corrugated fastener (tie) to each stud so it will fit into the mortar joint between the second and third courses of masonry. Bend them into place. See **Figure 62-2**.   Completed ☐

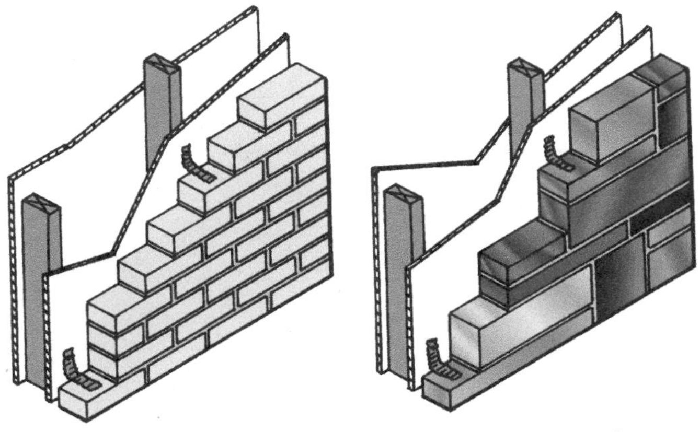

Brick veneer on frame   **Stone veneer on frame**

Goodheart-Willcox Publisher

Figure 62-2. Brick and stone veneer attached to frame construction.

6. String a bed of mortar for the third course, making sure to put mortar above and below the metal ties. Lay the third course.   Completed ☐

7. Continue laying courses until you reach the eighth course. Attach metal ties between the eighth and ninth courses.   Completed ☐

8. Lay the ninth and final course. Strike the joints and brush the wall.   Completed ☐

9. Inspect your work, clean up the area, and return tools and materials to their assigned places.   Completed ☐

Instructor's Initials: _____

Date: _____

Job 62 Review

After completing this job successfully, answer the following questions.

1. Brick, stone, or concrete masonry units are generally used in loadbearing applications. Why would they be used as a facing in an anchored veneered wall?

   _____

   _____

2. How are the masonry units bonded to the backup wall in anchored veneered walls?

   _____

3. Why is an air space of 1″ generally used in an anchored veneered wall?

   _____

   _____

4. How is the moisture that collects in the air space removed?

   _____

   _____

5. What kind of material is generally used as flashing in an anchored veneered wall?

_____

_____

_____

_____

Score: _____

Name _____ Date _____ Class _____

JOB 63: Building a 12" Composite Brick and Block Wall

OBJECTIVE: After you have completed this job, you will be able to build a 12" composite brick and block wall using the proper technique.

TEXTBOOK REFERENCE: Study the appropriate section of Chapter 16, *Wall Systems,* before starting this job.

Equipment

To accomplish this job, you will need the following tools and materials:

- Mason's tools
- Mortar
- Supply of bricks and blocks
- Supply of header blocks
- Hard hat, safety glasses, and other appropriate safety equipment

Recommended Procedure

A 12" composite wall is constructed in a manner similar to an 8" composite wall, but a header block is used in a 12" wall. The header block may be laid with the recessed notch up or down, depending on construction requirements. In **Figure 63-1**, header blocks were used for the sixth course bonding.

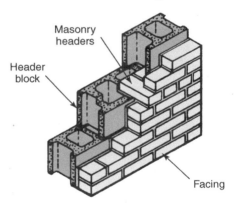

Goodheart-Willcox Publisher

Figure 63-1. A 12" composite wall with a masonry bond used to tie the wythes together.

The demonstration wall to be built will be 12" thick, 24" high, and 48" long. Header block will be used to provide sixth course bonding.

1. Assemble the tools and materials needed for this job and prepare a work area.   Completed ☐

2. Snap a chalk line to fix the location of the wall. Lay down a mortar bed for the course of concrete block. Lay three stretcher blocks as the back wythe.   Completed ☐

3. Lay three courses of bricks as the front wythe. Fill the collar joints and use full head joints. Be sure the bricks and blocks are the same height at this point.   Completed ☐

4. Lay the second course of blocks on the back wythe. These are header blocks with the notch facing up.   Completed ☐

5. Lay two courses of bricks on the front wythe in running bond. Then lay the next course as headers to tie the wythes together. Level, plumb, and straightedge the assembly.   Completed ☐

Copyright Goodheart-Willcox Co., Inc.
May not be reproduced or posted to a publicly accessible website.

6. Lay the third course of concrete blocks on the back wythe. Then lay bricks up to the height of the blocks (nine courses). Completed ☐

7. Tool the joints and brush the wall. Inspect your work. Completed ☐

8. Clean up the area and return the tools and materials to their assigned places. Completed ☐

Instructor's Initials: _____

Date: _____

Job 63 Review

After completing this job successfully, answer the following questions.

1. What is the basic difference in construction between an 8″ and a 12″ composite wall?
   _____
   _____

2. Why are no metal ties required in the 12″ composite wall?
   _____
   _____

3. Which wythe was begun first in this wall?
   _____

4. How many courses of bricks equaled one course of blocks in this job?
   _____

5. What is the name of the bond on the front wythe (brick)?
   _____

Score: _____

Name _____ Date _____ Class _____

JOB 64: Building Reinforced Masonry Walls

OBJECTIVE: After you have completed this job, you will be able to build a reinforced concrete block wall and a reinforced brick masonry wall using the proper technique.

TEXTBOOK REFERENCE: Study the appropriate section of Chapter 16, *Wall Systems,* before starting this job.

Equipment
To accomplish this job, you will need the following tools and materials:
- Mason's tools
- Chalk line
- Mortar
- Supply of bricks
- Supply of two-cell blocks and lintel blocks
- Two steel bars 3/8″ × 32″
- Four steel bars 1/2″ × 32″
- Hard hat, safety glasses, and other appropriate safety equipment

Recommended Procedure
Reinforced walls are built with steel reinforcement embedded with the masonry units. The walls are structurally bonded by grout, which is poured into the cavity (collar joint) between the wythes of masonry when solid units are used. Grout is poured in the cells of hollow units when concrete blocks are used.

Reinforced masonry walls should be reinforced with an area of steel not less than 0.002 times the cross-sectional area of the wall. Not more than 2/3 of this area may be used in either direction. Maximum spacing of principal reinforcement should not exceed 48″.

This job will provide the opportunity to build two reinforced masonry walls—one with hollow units (concrete block) and one with solid units (brick). Each demonstration wall will be 32″ long and 24″ high.

Reinforced Masonry Wall Using Hollow Units (Concrete Blocks)

1. Assemble the tools and materials needed to build this demonstration wall. Arrange your work area and snap a chalk line on the floor to maintain the location of the wall. Lay down a full mortar bed for the concrete blocks. Lay the first course of blocks as shown in **Figure 64-1**.

    Completed ☐

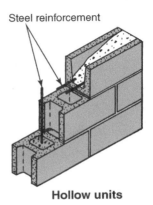

Goodheart-Willcox Publisher

Figure 64-1. Reinforced masonry wall using hollow masonry units.

2. Lay the second course of blocks as shown in **Figure 64-1**. Fill the exposed cell in the blocks on the first course and insert a 1/2″-diameter rebar. Completed ☐

3. Lay the top course using a lintel block. Place two 3/8″ diameter pieces of rebar in the lintel block. Add some mortar to hold the bars in place. Use a board across the end to hold back the mortar. Support it with two concrete blocks. Completed ☐

4. Strike the joints and brush the blocks. Inspect your work. Completed ☐

Reinforced Masonry Wall Using Solid Units (Brick)

1. Assemble the tools and materials needed to build this demonstration wall. Arrange your work area and snap a chalk line on the floor to maintain the location of the wall. Lay down a full mortar bed for the back wythe of bricks. Lay the back wythe as shown in **Figure 64-2**. Completed ☐

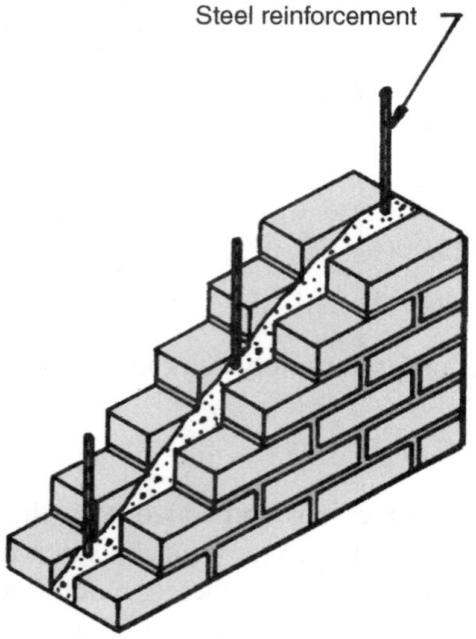

Figure 64-2. Reinforced masonry wall using solid units.

Goodheart-Willcox Publisher

2. Lay the front wythe of brick exactly as you laid the back wythe. Tool the joints and brush the wall. Completed ☐
3. Fill the void with mortar as shown in **Figure 64-2**. Place a board across the end to hold the mortar. Support the board with a concrete block. Completed ☐
4. Insert three 1/2″-diameter rebars vertically in the space between the wythes. Completed ☐
5. Examine your work, clean up the area, and return the tools and materials to their assigned places. Completed ☐

Instructor's Initials: _____

Date: _____

Job 64 Review

After completing this job successfully, answer the following questions.

1. How is grout used in a reinforced masonry wall when bricks are the masonry units?

_____

_____

Name _____

2. How is grout used in a reinforced masonry wall when hollow concrete blocks are the masonry units?
_____

3. What is the minimum amount of steel reinforcing that must be used in a reinforced masonry wall?
_____
_____

4. What size vertical and horizontal rebar was used in the concrete block wall?
_____

5. No horizontal reinforcing was used in the wall that was made from brick. Is horizontal reinforcing sometimes used in this wall type? Why?
_____
_____
_____

Score: _____

Notes

Name _____ Date _____ Class _____

JOB 65: Installing Steel and Concrete Reinforced Lintels

OBJECTIVE: After you have completed this job, you will be able to install steel lintels in brick veneer and concrete reinforced lintels in concrete block walls using the proper technique.

TEXTBOOK REFERENCE: Study the appropriate section of Chapter 16, *Wall Systems,* before starting this job.

Equipment

To complete this job, you will need the following tools and materials:

- Mason's tools
- Chalk line
- Mortar
- Supply of bricks
- Supply of blocks
- Steel angle 3″ × 3″ × 1/4″ by 56″ long
- Concrete reinforced lintel 7 5/8″ × 7 5/8″ × 64″
- Two thin sheet metal shims 6″ × 8″
- Hard hat, safety glasses, and other appropriate safety equipment

Recommended Procedure

Masonry above an opening must be supported with a lintel. Steel angles are generally used in brick and stone veneer walls. Precast concrete lintels or lintels made from lintel blocks are frequently used in concrete block walls. Whatever the type or application, the lintel should be stiff enough to resist bending in excess of 1/360 of the span.

In this job you will install a steel angle in a brick veneer wall and a precast concrete lintel in a concrete block wall.

Steel Angle in a Brick Veneer Wall

1. Assemble the tools and materials needed to build two sections of a single wythe of bricks, each three bricks in length with a space between them of 48″. The total assembly will be five courses high. ☐ Completed

2. Snap a chalk line to locate the face of the wythe. Locate the center of the chalk line and measure off 24″ in each direction along the line. This 48″ distance is the length of the opening in the wall. ☐ Completed

3. Lay three bricks on either side of the "opening." Lay a second course in running bond on the first course. Lay a third course to bring the wall segments up to 8″ in height. ☐ Completed

4. Strike the joints. Carefully set the steel angle across the openings so it is supported 4″ on either end. ☐ Completed

5. Trim the sharp edge off several bricks along the edge that will sit in the radius on the steel angle. Lay the fourth course of bricks along the entire wall. ☐ Completed

6. Lay the fifth course of bricks to complete the wall. Strike the mortar joints and brush the wall. ☐ Completed

Precast Concrete Lintel in a Concrete Block Wall

1. Assemble the tools and materials needed to build two sections of an 8″ concrete block wall. Each section will be 24″ long with a 48″ space between. The total assembly will be three courses high. ☐ Completed

2. Snap a chalk line to locate the face of the wall. Locate the center of the chalk line and measure off 24″ in each direction along the line. This 48″ distance will be the length of the opening in the wall. Completed ☐

3. Lay one block and a half block on either side of the "opening." Lay a second course in running bond. Strike the joints. Let the mortar cure for several hours before proceeding to the next step. Completed ☐

4. Place a sheet metal shim on top of the blocks nearest the opening. This will provide support for the concrete lintel. Place the mortar, then ask someone to help you set the lintel into place. It should be 8″ longer on either end than the opening. Completed ☐

5. Check the lintel to be sure it is level and positioned properly. Complete the course. Strike the joints and brush the wall. Completed ☐

6. Inspect your work, clean up the area, and return all tools and materials to their assigned places. Completed ☐

Instructor's Initials: _____

Date: _____

Job 65 Review

After completing this job successfully, answer the following questions.

1. What is the function of a lintel?

   _____
   _____

2. What kind of lintels are frequently used in concrete block walls?

   _____
   _____

3. What type of lintel is generally used in brick and stone veneer walls?

   _____

4. What is the maximum bending allowed in lintels regardless of their type?

   _____

5. What is the recommended length of support at either end of a steel angle lintel?

   _____

6. What is the function of the sheet metal skin at either end of the concrete lintel?

   _____
   _____

Score: _____

Name _____ Date _____ Class _____

JOB 66: Building Reinforced Concrete Block and Brick Lintels

OBJECTIVE: After you have completed this job, you will be able to build reinforced concrete block and brick lintels using the proper technique.

TEXTBOOK REFERENCE: Study the appropriate section of Chapter 16, *Wall Systems,* before starting this job.

Equipment

To complete this job, you will need the following tools and materials:

- Mason's tools
- Mortar
- Supply of lintel blocks
- Supply of bricks
- Supply of reinforcing steel bars
- 4 pieces of plywood, 12″ × 16″
- Hard hat, safety glasses, and other appropriate safety equipment

Recommended Procedure

In reinforced masonry lintels, the steel is completely protected from the elements. The initial cost is less than steel angle lintels since less steel is required.

In this job, you will build a reinforced brick lintel and a reinforced concrete block lintel. The products of this job can be used for load testing and experimentation after they have cured for 28 days.

Reinforced Concrete Block Lintel

1. Assemble the tools and materials required to build a reinforced concrete block lintel that is 8″ × 8″ × 48″. Completed ☐

2. Lay three concrete lintel blocks end to end on the floor with mortar between each. Tool the joints. **Figure 66-1** shows a section of the lintel. Completed ☐

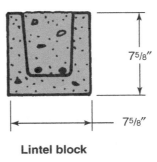

Lintel block

Goodheart-Willcox Publisher

Figure 66-1. Section of a reinforced concrete block lintel.

3. Place a piece of plywood at each end of the assembly, with a block against the boards to prevent the mortar from leaking out. Completed ☐

4. Place a mortar bed approximately 1/2″ thick in the lintel blocks to ensure that the rebars are completely embedded. Lay in two pieces of rebar 3/8″ diameter by 48″ long. Refer to **Figure 66-1**. Completed ☐

5. Completely fill the lintel block cavity with mortar or grout. Work it into all crevices. Smooth the top.   Completed ☐

6. Allow the assembly to cure for several days before moving it. Use the lintel for load testing.   Completed ☐

Reinforced Brick Masonry Lintel

1. Assemble the tools and materials required to build a reinforced brick lintel that is 8″ wide by five courses high by 48″ long.   Completed ☐

2. Lay the bed courses of brick on the floor with full head and collar joints. Do not use a bed joint, since the assembly will be moved later. Refer to **Figure 66-2**.   Completed ☐

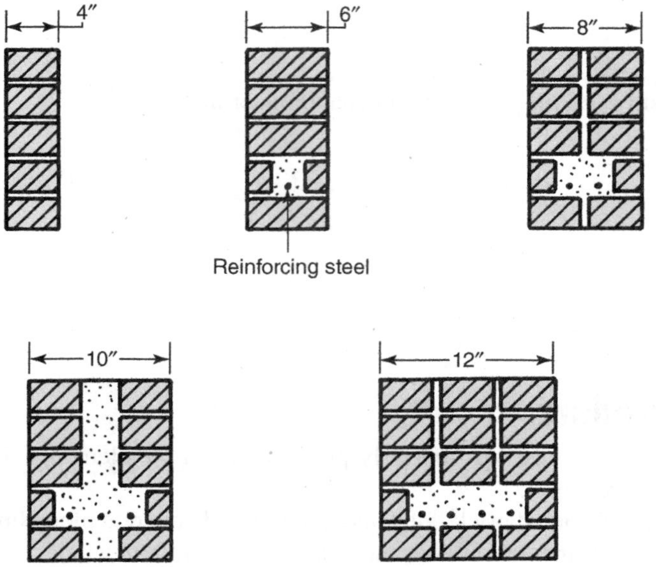

Figure 66-2. Reinforced brick masonry lintels.

3. Cut several bricks lengthwise for the second course of brick. Lay the second course of queen closures. Place a piece of plywood at each end to contain the grouted cavity.   Completed ☐

4. Lay a mortar bed approximately 1″ thick between the wythes. Work it into the cracks and crevices. Place two 3/8″ rebars as shown in **Figure 66-2** for the 8″ lintel. The rebar should extend the full length of the lintel.   Completed ☐

5. Fill the cavity with mortar and work it into all spaces and voids. Smooth off the top.   Completed ☐

6. Lay the next three courses with full bed, head, and collar joints. The total assembly must act as a single unit to be effective. Tool the joints and brush the bricks.   Completed ☐

7. Do not move the lintel for several days. It may then be used for load tests.   Completed ☐

8. Inspect your work, clean up the area, and return all tools and materials to their assigned places.   Completed ☐

Instructor's Initials: _____

Date: _____

Name _____

Job 66 Review

After completing this job successfully, answer the following questions.

1. Why are reinforced brick lintels less expensive than steel angle lintels?

   _____
   _____

2. Why are special lintel blocks needed to build a concrete block lintel?

   _____
   _____

3. How many days are required for concrete or mortar to cure?

   _____

4. What was the most time-consuming operation in building a brick lintel?

   _____

5. What is the most likely reason for failure of a brick lintel?

   _____
   _____

Score: _____

Notes

Name _____ Date _____ Class _____

JOB 67: Building Masonry Sills and Installing Stone Sills

OBJECTIVE: After you have completed this job, you will be able to build masonry sills and install stone sills using the proper technique.

TEXTBOOK REFERENCE: Study the appropriate section of Chapter 16, *Wall Systems,* before starting this job.

Equipment

To complete this job, you will need the following tools and materials:

- Mason's tools
- Mortar
- Supply of bricks
- Flashing
- One stone sill
- Section of frame wall with opening
- Hard hat, safety glasses, and other appropriate safety equipment

Recommended Procedure

The primary function of a sill is to channel water away from the building. A sill may consist of a single unit or multiple units. It may be built in place or prefabricated. Sills are made from a variety of materials—brick, stone, concrete, or metal. Single units are either slip sills or lug sills. See **Figure 67-1**.

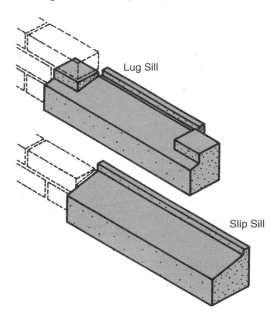

Goodheart-Willcox Publisher

Figure 67-1. A one-piece lug and slip sill.

A slip sill may be left out when the opening is being laid and set at a later time, but a lug sill should be set when the masonry is up to the bottom of the opening. Brick sills may be installed during or after the face bricks are laid.

In this job, you will lay a brick sill using rowlock headers and install a stone slip sill in a brick veneered wall.

Brick Masonry Sill

1. Assemble the tools and materials needed to lay a short veneered wall and install a brick masonry sill. Study **Figure 67-2** to see how a brick sill is constructed. Completed ☐

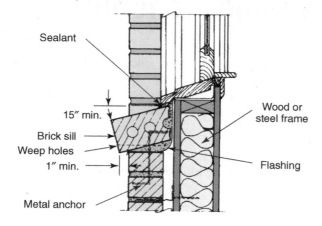

Figure 67-2. Brick sill in a veneered wall on frame backup.

2. Using a frame wall backup, lay the veneer up to three courses above the bottom of the window opening. Completed ☐

3. Set bricks on edge in the opening where the sill will be built to determine proper spacing. Make note of the mortar joint thickness required. Thinner mortar joints are preferred over thick ones. Completed ☐

4. Cut enough bricks to build the sill. Refer to the drawing again. Completed ☐

5. Install flashing to shed water and prevent mortar from falling into the air space. Apply a generous mortar bed and lay the rowlock headers. Be sure the head joints are full. Completed ☐

6. Tool the joints and brush the bricks. Inspect your work to be sure it meets the desired standards. Completed ☐

Stone Slip Sill

1. Assemble the tools and materials needed to install a slip sill in a brick veneered wall on frame backup. Study **Figure 67-3** to see how a stone slip sill is positioned in the wall. Completed ☐

Figure 67-3. Stone slip sill in a brick veneered wall on frame backup.

Name _____

2. Using a frame wall backup, lay the veneer up to three courses below the window opening. Measure the sill to determine the desired height of masonry to mount the sill at the proper height. Masonry units may have to be cut so the spacing works out right. Prior planning is important.   Completed ☐

3. Complete the veneer to the desired height to set the sill. Set the sill in a full mortar bed. Be sure it is level and positioned properly with respect to the bottom of the window.   Completed ☐

4. Continue three more courses of veneer on either side of the window. Strike the joints and brush the wall.   Completed ☐

5. Examine your work, clean up the area, and return all tools and materials to their assigned places.   Completed ☐

Instructor's Initials: _____

Date: _____

Job 67 Review

After completing this job successfully, answer the following questions.

1. What is the function of a sill?
   _____
   _____

2. What three materials are used most frequently for sills in masonry construction?
   _____

3. What are the basic differences between a slip sill and a lug sill?
   _____
   _____
   _____
   _____
   _____
   _____

4. A brick sill is neither a slip nor a lug sill but is composed of several units. When is the appropriate time to build a brick sill?
   _____
   _____
   _____

5. What position are bricks usually laid in when a sill is being formed?
   _____
   _____

6. When planning for a stone sill, what is the critical height dimension location?
   _____
   _____

Score: _____

Notes

Name _____ Date _____ Class _____

JOB 68: Building a Brick Masonry Arch

OBJECTIVE: After you have completed this job, you will be able to build a brick masonry arch using the proper technique.

TEXTBOOK REFERENCE: Study the appropriate section of Chapter 16, *Wall Systems,* before starting this job.

Equipment

To complete this job, you will need the following tools and materials:
- Mason's tools
- Chalk line
- Mortar
- Supply of bricks
- Supply of 8-penny nails
- Construction lumber to build support centering
- Hard hat, safety glasses, and other appropriate safety equipment

Recommended Procedure

An arch is normally classified by the curve of its intrados (the curve that bounds the lower edge of the arch) and by its function, shape, or architectural style. Arches are classified as major and minor arches. Minor arches have spans that do not exceed 6′ with a maximum rise-to-span ratio of 0.15.

Jack and segmental arches are common types. In this job, you will build a segmental arch in an 8″ solid brick wall. The arch will use the centering built for Job 56. The arch will be 36″ wide with a rise of 3″. See **Figure 68-1**.

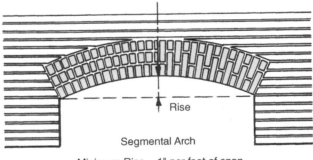

Segmental Arch

Minimum Rise = 1″ per foot of span

Goodheart-Willcox Publisher

Figure 68-1. Typical segmental arch.

1. Assemble the tools and materials needed for this job and arrange a work area. Examine the centering already constructed or construct your own following the procedure described in Job 56.   Completed ☐

2. Snap a chalk line to establish a wall 6′ long and 8″ thick. The 36″-wide arch will be centered in the wall. Mark the location of the arch in the wall.   Completed ☐

3. Lay five courses of bricks on each side of the arch. Make sure these bricks have full bed and head joints, because they will support the weight of the masonry above the arch.   Completed ☐

4. Measure the distance from the floor to the top of the fifth (top) course of bricks. Subtract from that dimension the height of the centering at the end. Cut four pieces of 2″ × 4″ lumber to support the ends of the centering (two under each end).   Completed ☐

5. Tack the support pieces to the rib boards. You will need to pull these nails when you remove the centering. Check the assembly to be sure it is plumb and steady. You may have to shim one or more legs to produce a solid platform for work. Completed ☐

6. Lay the centering on its side and place bricks in a soldier course around the arch curve to determine the proper spacing. The bricks should have very narrow mortar joints at the bottom, but not touch. Mark the position of each brick on the form. Completed ☐

7. Insert the assembled centering and support legs between the two wall sections. Be careful not to disturb them. If you do, take them down and begin again. Completed ☐

8. Lay the back wythe first. Place one or two bricks on the left and then on the right side. Work from both sides toward the middle. Completed ☐

9. Follow the same procedure in laying the bricks on the front wythe. Be sure to fill the collar joint. Try not to let mortar squeeze out too much on the bottom side of the units. These joints must be pointed later. Completed ☐

10. Tool the joints and check the appearance of the arch. If it is not uniform and clean, tear it down and begin again. Completed ☐

11. Complete the wall until it reaches a height that is two courses above the highest point of the arch. Notice that you will have to cut several bricks to fit the curve. Measure twice and cut once. This is the most difficult part of the job. Completed ☐

12. Tool the joints and brush the wall. Completed ☐

13. When the mortar has cured a few days, remove the shims and nails from the supports from beneath the centering, then carefully remove the centering. Get someone to help you. Do not let the centering fall. Completed ☐

14. Remove excess mortar from the bottom of the arch and point any joints that need it. Completed ☐

Instructor's Initials: _____

Date: _____

Job 68 Review

After completing this job successfully, answer the following questions.

1. Identify four bases for the classification of arches.

   _____
   _____

2. Most arches used in residential or commercial construction today are minor arches. What is a minor arch?

   _____
   _____
   _____

3. If someone has not already prepared a pattern of the arch curve for you, how could you scribe the curve on a piece of cardboard using tools and materials typically found in a mason's toolbag?

   _____
   _____
   _____

Name _____

4. Why is the centering for the arch built 1″ less than the thickness of the wall?

   _____

   _____

   _____

5. What is the purpose of the keystone?

   _____

   _____

   _____

Score: _____

Notes

Name _____ Date _____ Class _____

JOB 69: Forming Movement Joints in Concrete and Masonry

OBJECTIVE: After you have completed this job, you will be able to form movement joints in concrete and masonry using proper technique.

TEXTBOOK REFERENCE: Study the appropriate section of Chapter 16, *Wall Systems,* before starting this job.

Equipment

To complete this job, you will need the following tools and materials:

- Mason's tools
- Mortar
- Supply of bricks
- Supply of concrete blocks
- Elastic expansion joint materials
- Concrete
- Form material (2″ × 4″)
- Shovel
- Control joint blocks
- Hard hat, safety glasses, and other appropriate safety equipment

Recommended Procedure

Because all materials in a building experience changes in volume, a system of movement joints is necessary to allow movement to occur. Failure to permit these movements may result in cracks in masonry construction. Types of movement joints in buildings include expansion joints, control joints, building expansion joints, and construction joints. Each type of movement joint is designed to perform a specific task; they should not be used interchangeably.

Expansion Joint

An expansion joint is used to separate brick masonry into segments to prevent cracking due to changes in temperature, moisture expansion, elastic deformation due to loads, and creep. Expansion joints may be horizontal or vertical. The joints are formed of highly elastic materials placed in a continuous, unobstructed opening, through each wythe. See **Figure 69-1**.

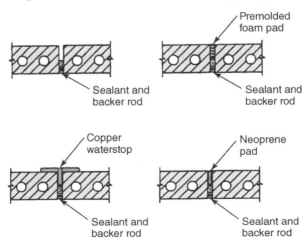

Brick Industry Association

Figure 69-1. Typical expansion joints in brick masonry.

1. Lay a short, single-wythe wall six units long in running bond, but include a vertical expansion joint in the center of the wall. The joint should be the same width as a regular mortar joint. Some bricks will have to be cut.   Completed ☐

2. Apply an elastic expansion joint material to the joint. Follow the manufacturer's instructions.   Completed ☐

Control Joint in Concrete

A control joint in concrete or concrete masonry creates a plane of weakness that, when used in conjunction with reinforcement or joint reinforcement, controls the location of cracks due to volume changes resulting from shrinkage and creep. A control joint may be made of inelastic materials, and will open rather than close.

1. Build a form from 2″ × 4″ lumber to cast a flat slab 8′ long by 1′ wide by 3 1/2″ thick on a level surface. Form an anchor in the soil at each end by digging a hole in the ground 12″ to 14″ deep.   Completed ☐

2. Fill the form with concrete and finish the slab in the normal manner. Cut a control joint at the midpoint.   Completed ☐

3. Observe the control joint periodically to note any separation at the control joint.   Completed ☐

Control Joint in Concrete Masonry

A special concrete masonry unit is available that controls volume change in a wall. Placed at specific intervals, this block relieves stress in the wall. See **Figure 69-2**.

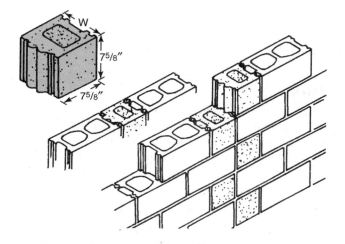

Goodheart-Willcox Publisher

Figure 69-2. Special concrete block designed to relieve stresses in a wall.

1. Lay a concrete block wall 48″ long with a vertical control joint at the center point of the wall. Use regular 8″ × 8″ × 16″ stretchers and 8″ × 8″ × 8″ control joint blocks as shown in **Figure 69-2**. Lay the wall three courses high.   Completed ☐

Name _____

2. Rake out the mortar along the control joint and fill with elastic expansion joint material. Tool the other joints and brush the wall. **Figure 69-3** shows a control joint in masonry construction.

Completed

Figure 69-3. A typical building control joint.

Instructor's Initials: _____

Date: _____

Job 69 Review

After completing this job successfully, answer the following questions.

1. Why are movement joints necessary in concrete and masonry structures?

   _____
   _____
   _____

2. Name several types of movement joints that may be used in a building.

   _____
   _____
   _____

3. What kind of material is placed in an expansion joint?

   _____

4. How does a control joint in a concrete slab work?

   _____
   _____
   _____

Score: _____

Notes

Name _____ Date _____ Class _____

JOB 70: Installing Masonry Pavers on a Rigid Base

OBJECTIVE: After you have completed this job, you will be able to install masonry pavers on a rigid base using the proper technique.

TEXTBOOK REFERENCE: Study the appropriate section of Chapter 17, *Paving and Masonry Construction Details,* before starting this job.

Equipment

To complete this job, you will need the following tools and materials:
- Mason's tools
- Mortar
- Supply of pavers
- Existing concrete slab
- Masonry saw
- Hard hat, safety glasses, and other appropriate safety equipment

Recommended Procedure

Masonry paving systems can be installed with mortar (rigid) or without mortar (flexible or *hand-tight*). Most standard pavers used with mortar allow for a 3/8" mortar joint. Rigid systems must be installed on a rigid base. A concrete slab usually provides this base. See **Figure 70-1**.

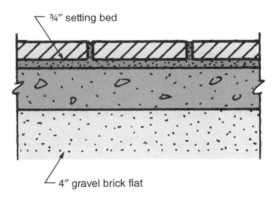

Goodheart-Willcox Publisher

Figure 70-1. Section through pavers on concrete slab.

This job will provide an opportunity to pave a 6′ square area on a concrete slab.

1. If an existing concrete slab is not available, then form and pour a 6′ square slab for this job. Assemble the necessary tools and materials to install pavers on the slab. Completed ☐

2. Select any concrete, brick, or clay paver that you like. Choose an appropriate pattern bond and place the pavers on the slab without mortar to determine the best arrangement. Some pavers may need to be cut to center the pattern on the slab. Completed ☐

3. When the arrangement and spacing have been decided upon, mix the mortar. Masonry cement is often not recommended by manufacturers of pavers. Completed ☐

4. Provide a solid bed joint of mortar about 3/4" thick and lay the pavers along one side. Be sure the spacing is uniform. If pavers need to be cut, use the masonry saw. Completed ☐

5. Lay the pavers along adjacent sides. Then fill in the remaining space working from these two sides.   Completed ☐

6. When moving around on the pavers while the mortar is wet, use large pieces of 3/4" plywood to spread the load. Tool the joints and clean off any mortar on the pavers with a damp sponge. Do not sponge the mortar joints.   Completed ☐

7. Inspect your work, clean up the area, and return all tools and materials to their assigned places.   Completed ☐

Instructor's Initials: _____

Date: _____

Job 70 Review

After completing this job successfully, answer the following questions.

1. Masonry pavers may be placed on a rigid or flexible base. What primary factor should be considered in deciding which to use?

   _____
   _____
   _____

2. Why must a rigid base be used if the pavers are to have mortar between the joints?

   _____
   _____
   _____

3. How can fresh mortar be removed from pavers?

   _____
   _____
   _____

4. For pavers laid on a rigid base, what is the recommendation for bed joint thickness?

   _____
   _____

5. The base is important for a flexible installation. Describe the recommended type of base for hand-tight units.

   _____
   _____
   _____

Score: _____

Name _____ Date _____ Class _____

JOB 71: Building Concrete and Masonry Steps

OBJECTIVE: After you have completed this job, you will be able to build concrete and masonry steps using the proper technique.

TEXTBOOK REFERENCE: Study the appropriate section of Chapter 17, *Paving and Masonry Construction Details*, before starting this job.

Equipment

To complete this job, you will need the following tools and materials:
- Mason's tools
- Materials for concrete
- Concrete mixer
- Form materials
- Bricks or pavers
- Shovel
- Concrete finishing tools
- A sloping site
- Hard hat, safety glasses, and other appropriate safety equipment
- Sketch pad

Recommended Procedure

This job will provide an opportunity to build a simple set of steps similar to those shown in **Figure 71-1**. This is a good class project.

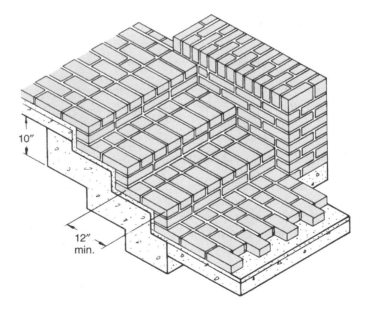

Goodheart-Willcox Publisher

Figure 71-1. Brick steps laid over a concrete foundation.

1. Assemble the tools and materials needed to prepare the foundation excavation and forms for the concrete base for a set of steps similar to those shown in **Figure 71-1**. Completed ☐

2. On a sketch pad, record the drop in elevation where you plan to build the steps. Plan the steps around this information. Follow good design principles. For example, the ideal riser height is about 7″, while the ideal tread width is about 12″. Plan the stairs in such a way that masonry unit sizes are also taken into consideration. Ask your instructor to review your plan.   Completed ☐

3. When the plan is complete and acceptable, begin excavation for the foundation and build the forms. Check the dimensions several times to be sure you have followed your plan.   Completed ☐

4. Mix the concrete and place it in the forms. Rod or vibrate the concrete to be sure all voids are filled. Leave the surface slightly rough so the mortar will stick to it.   Completed ☐

5. When the concrete has cured for several days, remove the forms and examine your work. Lay out one step without mortar to check the spacing. Then, mix some mortar and lay the units on the lowest level first. Work your way up the steps.   Completed ☐

6. Be careful not to bump any of the previously laid units. Tool the joints as the mortar hardens to the proper degree. Brush the bricks. Use a damp sponge to remove any mortar smears on the bricks. Do not sponge the mortar joints.   Completed ☐

7. Build the walls on either side of the steps using any pattern you choose.   Completed ☐

8. Tool all mortar joints and brush the bricks. Check over the whole job and admire your project. Congratulations, you designed the project and built it "from the ground up"!   Completed ☐

9. Smooth the grade around the steps and return all tools and materials to their assigned places.   Completed ☐

Instructor's Initials: _____

Date: _____

Job 71 Review

After completing this job successfully, answer the following questions.

1. When planning a set of steps, why is it necessary to know the total rise of the steps?

   _____
   _____
   _____

2. What is the ideal riser height of a step?

   _____

3. As the riser height increases from 7″ to 9″, what should happen to the tread width?

   _____
   _____

4. Why is building a good solid foundation necessary for a set of steps?

   _____
   _____
   _____

5. What type of bricks should be used for steps or paving in climates that experience freezing?

   _____

Score: _____

Name _____ Date _____ Class _____

JOB 72 / Building a Masonry Fireplace and Chimney

OBJECTIVE: After you have completed this job, you will be able to build a masonry fireplace and chimney using the proper technique.

TEXTBOOK REFERENCE: Study the appropriate section of Chapter 17, *Paving and Masonry Construction Details,* before starting this job.

Equipment

To complete this job, you will need the following tools and materials:
- Mason's tools
- Mortar
- Supply of bricks
- Firebricks
- Damper
- Lintel (angle steel)
- Supply of concrete blocks
- Fire clay
- Flue liners
- Flue blocks
- Pad of graph paper
- Hard hat, safety glasses, and other appropriate safety equipment

Recommended Procedure

This job should be a class project in a realistic setting—a home under construction. If this is not possible, the project can be performed in the lab. Rather than having every detail planned for you, in this job you will do the planning. The basic operation will be presented to keep you on track. Study the section on fireplace and chimney construction in your text before starting this job to become familiar with the parts of a fireplace and the terminology.

1. The most popular and functional fireplace for most residential settings is a single-face fireplace with an opening about 36″ wide and 29″ high. Turn to **Figure 17-14** in your text, which shows specifications for fireplace opening height, hearth size, and flue size. Note that a 36″ wide × 29″ high opening requires a 16″ hearth and 12″ × 12″ flue liners. The damper for this fireplace is 40″ on the outside of the front flange but tapers back to 32 3/4″ on the back. Continue studying the specifications until you feel sure you have the dimensions you need to build the fireplace.

Completed ☐

2. Study **Figure 72-1**, which shows the structure from the footing to the chimney cap. The drawing shows every brick. When you can visualize building the entire structure, you are ready to begin the construction.

Completed

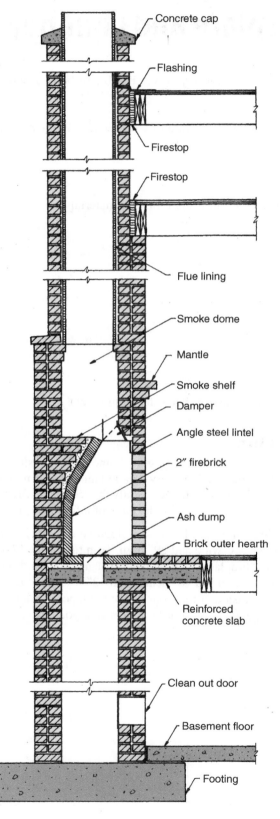

Figure 72-1. Fireplace and chimney components.

Name _____

3. Sketch a plan view of the fireplace on graph paper to scale. Make the drawing large enough so details can be represented. A scale of 1/8″ = 1″ would fit on 8 1/2″ × 11″ paper if it were turned sideways. Include brick sizes in your plan. You may have to do some research. For example, what are the dimensions of a firebrick? What size mortar joint is used with firebrick? What is fire clay? What are the code requirements in your area related to masonry fireplaces? These are good class discussion topics. Complete your plan. Completed ☐

4. The foundation for a chimney is larger and thicker than a typical foundation wall footing because the fireplace and chimney are very heavy and need a large footing. The fireplace/chimney should be a freestanding structure. Framing should not be closer than 2″ to the chimney. Build the foundation for your fireplace using concrete blocks, bricks, or a combination. Completed ☐

5. Be careful as you approach the floor level. You should plan the elevation (height) of the outer and inner hearth. Will it be raised or level with the finished floor as **Figure 72-1** shows? Plan it the way you want it to be and then work your plan. Formerly, the outer hearth was supported by corbeling out the bricks, but that method is seldom used anymore. You could corbel out the bricks but probably should plan for a reinforced slab. Complete the job to the floor level. Completed ☐

6. Install the bricks on the outer hearth last because it will be covered in mortar by the time the job is finished. Place the firebricks on the inner hearth. These units should be arranged with the ash dump as the central feature. Locate the placement of the ash dump and then arrange the firebricks. Lay them on a thick mortar bed with fire clay between them. Read the instructions on the can or bucket. Let the bricks on the hearth run beyond the inside dimensions of the firebox to support the firebricks that line the firebox. Completed ☐

7. Bring up the sides of the firebox and chimney to the level where the back begins to slope forward. This is where the job gets tricky. You must cut the firebrick to fit the angles formed by the tapered sides and sloping back. Be sure to fill in the space behind and beside the liner with mortar and pieces of bricks or blocks. Prop up the sloping firebricks as you bring them up to the level of the damper. Completed ☐

8. Once you reach the height of the front opening (29″), install the 3″ × 4″ × 1/4″ angle steel. Trim off the back edge of the bricks so they fit properly against the angle steel lintel. Set the angle directly on the course of bricks beneath it so it can move as it heats up and cools down. Lay a couple of courses above the opening and then install the damper. If you calculated accurately, the damper will cover the opening. If not, you will have to tear it out and start again. Completed ☐

9. Continue building the chimney until you reach the mantle level. You do not have to have a mantle, but at least consider it while you can. Corbel the bricks at the top of the smoke chamber until the opening is the size of the inside dimensions of the flue liner. Wait until the mortar has hardened somewhat before setting the first flue liner on its supports. Completed ☐

10. Set the first flue liner and support it on all sides as you lay up the chimney. Frequently, the width of the chimney is racked back once the masonry clears the height of the fireplace. This is an option, but not a necessity. Completed ☐

11. Continue laying up the chimney and flue liners until the desired height is reached. Build a form around the top of the chimney and pour the chimney cap. The top flue liner should protrude a couple of inches above the top of the cap. Completed ☐

12. Complete the outer hearth. Tool the joints and brush the bricks. Be proud of your work and strive to do it better each time. Completed ☐

Instructor's Initials: _____

Date: _____

Job 72 Review

After completing this job successfully, answer the following questions.

1. What effect does increasing the size of a fireplace opening have on the flue size and height of the chimney?

   _____

   _____

   _____

2. What part of a traditional masonry fireplace is most responsible for reflecting heat into the room?

   _____

3. Why is a fireplace/chimney a freestanding structure?

   _____

   _____

   _____

4. In some older fireplaces, regular bricks were used to line the firebox. Why is that not a recommended practice today?

   _____

   _____

   _____

5. What material are modern masonry flue liners made from?

   _____

6. If cutting a masonry flue liner becomes necessary, what tool should be used and why?

   _____

   _____

   _____

Score: _____

Name _____ Date _____ Class _____

JOB 73 / Building a Masonry Garden Wall with Coping

OBJECTIVE: After you have completed this job, you will be able to build a masonry garden wall with a coping using the proper technique.

TEXTBOOK REFERENCE: Study the appropriate section in Chapter 16, *Wall Systems*, before you begin this job.

Equipment

To complete this job, you will need the following tools and materials:

- Mason's tools
- Pad of graph paper
- Mortar
- Supply of bricks or blocks
- Stone or molded brick coping
- Concrete footing
- Hard hat, safety glasses, and other appropriate safety equipment

Recommended Procedure

Garden walls are freestanding structures that include perforated walls, straight walls, pier and panel walls, and serpentine walls. Each has its particular application and style.

In this job, you will build a panel-type garden wall of your own design. The specifications are that the wall should be 6′ in length, 4′ high, rest on a footing, terminate with a 12″ square post at each end, and be made from brick or block masonry.

1. Using graph paper, draw the garden wall panel that you plan to build. Your drawing should be accurate and complete, because the finished product will be compared to your drawing. Design your wall. Completed ☐

2. Show your design to your instructor and obtain approval before beginning the project. Completed ☐

3. Assemble the tools and materials needed for the project. Completed ☐

4. Build the wall using proper technique. Include reinforcing if it is necessary for your design. Refer to the text for recommendations to resist wind loads. See **Figure 73-1** for ideas for finishing off the wall at the top. Completed ☐

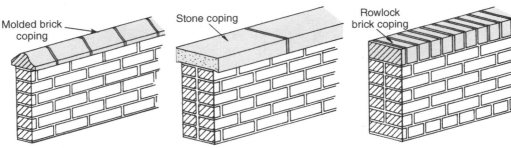

Goodheart-Willcox Publisher

Figure 73-1. Straight garden walls with three commonly used copings.

5. Discuss the strengths and weaknesses of your design with your instructor. Think about what you would do differently if you were building the wall a second time. Learn from your experience.

Completed ☐

Instructor's Initials: _____

Date: _____

Job 73 Review

After completing this job successfully, answer the following questions.

1. What are four types of garden walls?

   _____
   _____
   _____

2. Why is a footing or foundation that reaches below the frost line necessary for a quality garden wall?

   _____
   _____
   _____

3. Steel reinforcing might not be necessary in short, thick walls. How can you determine whether or not reinforcing should be included in a wall that you plan to build?

   _____
   _____
   _____

4. What is the function of the coping on a garden wall?

   _____
   _____
   _____

5. As a rule of thumb, what is the maximum height of an 8″-thick straight garden wall without reinforcing?

   _____

Score: _____

Name _____ Date _____ Class _____

JOB 74: Corbeling and Racking a Masonry Wall

OBJECTIVE: After you have completed this job, you will be able to build a masonry wall that includes corbeling and racking using the proper technique.

TEXTBOOK REFERENCE: Study the appropriate section in Chapter 16, *Wall Systems,* before you begin this job.

Equipment

To complete this job, you will need the following tools and materials:
- Pad of graph paper
- Mason's tools
- Mortar
- Supply of bricks
- Hard hat, safety glasses, and other appropriate safety equipment

Recommended Procedure

A corbel is defined as a shelf or ledge formed by projecting successive courses of masonry out from the face of the wall. Racking is defined as masonry in which successive courses are stepped back from the face of the wall.

Generally, the total horizontal projection should not exceed one-half the thickness of a solid wall, or one-half the thickness of the veneer of a veneered wall. A single course should not exceed one-half of the unit height or one-third of the unit bed depth, whichever is less.

From these limitations, the minimum slope of the corbeling can be established as 63 degrees and 26 minutes, measured from the horizontal to the face of the corbeled surface. There is no limitation on the distance each unit may be racked, as long as the unit cores are not exposed.

In this job, you will design a demonstration wall that includes a section of corbeled wall and a section of racked wall. The overall dimensions of the wall should be 12″ thick by 32″ long by twelve courses high. The structure should be stable.

1. Using a piece of graph paper, draw the plan view and elevations (side views) of your wall. Completed ☐
   Be sure to stay within the guidelines presented in the recommended procedure for this job.
   The rest is up to you.

2. Discuss your plan with your instructor before building it. Your instructor may offer suggestions. Completed ☐

3. Build your wall using your very best workmanship. Build something that you can be proud of! Completed ☐

Instructor's Initials: _____

Date: _____

Job 74 Review

After completing this job successfully, answer the following questions.

1. Define *corbel.*

_____

_____

_____

2. Describe *racking*.

   _____
   _____
   _____

3. The rules associated with corbeling are more stringent than for racking. Why?

   _____
   _____
   _____

4. How far out can a corbel be extended in a solid wall?

   _____

5. How far may a unit be racked?

   _____

   Score: _____

Name _____ Date _____ Class _____

JOB 75: Building a Mortarless Retaining Wall

OBJECTIVE: After you have completed this job, you will be able to build a mortarless retaining wall using the proper technique.

TEXTBOOK REFERENCE: Study the appropriate section in Chapter 16, *Wall Systems,* before you begin this job.

Equipment

To complete this job, you will need the following tools and materials:
- Level
- Shovel
- Crushed stone
- Sand
- Masonry units
- One dozen 1" square stakes about 12" long
- Mason's hammer
- Drain tile (optional)

Recommended Procedure

Concrete masonry units that incorporate a lip on the back side are popular for retaining walls, tree rings, planters, and edgings. The lip provides proper alignment and prevents forward wall movement. Units are produced in straight or tapered designs. Tapered units are especially designed to form curved walls. See **Figure 75-1**. Typical unit sizes include 6" × 16" × 12", 6" × 17 1/4" × 12", and 3" × 17" × 12". Sizes vary from one manufacturer to another.

Anchor Wall Systems, Inc.

Figure 75-1. Mortarless retaining wall using concrete masonry units.

In this job, you will build a 6'-diameter tree ring, two courses high. This is a good community project for the class.

1. Select an appropriate tree. Locate a 6'-diameter circle with the tree at the center. Mark the circle in some fashion.   Completed ☐

2. Remove some of the topsoil to below the grass level. Using the level, locate the desired height of the first course. Drive a stake in the ground to that height. Transfer the height to the next stake and continue around the circle until you reach the starting point.   Completed ☐

3. Remove more soil to allow for a base of crushed stone for the masonry units. Add the crushed stone to the desired height. The surface of the stone should be level.   Completed ☐

Copyright Goodheart-Willcox Co., Inc.
May not be reproduced or posted to a publicly accessible website.

4. Install the base course of tapered units. These units do not have a lip on the back. Level each unit and fill in around it with stone or sand. Remove the stakes as you proceed around the tree. Completed ☐

5. Complete the base course. Add a drain tile behind the wall, if necessary. Completed ☐

6. Install the second course. These units should have a lip on the back edge to ensure proper alignment. Stagger the joints. Complete this course. Completed ☐

7. Backfill with topsoil or sand. Do not pile soil up around the tree. Try to maintain the natural grade at the base of the tree. Completed ☐

8. Admire your work. Clean up the area and return all tools and materials to their assigned places. Completed ☐

Instructor's Initials: _____

Date: _____

Job 75 Review

After completing this job successfully, answer the following questions.

1. What feature is incorporated in all mortarless retaining walls?
   _____
   _____

2. Why is the foundation important in a mortarless retaining wall?
   _____
   _____
   _____

3. How could a construction level have been used to level the foundation for your tree ring project?
   _____
   _____
   _____
   _____

4. What two functions did the lip on the lower back edge of the concrete retaining wall units serve?
   _____
   _____
   _____

5. When would a drain tile be necessary in a retaining wall?
   _____
   _____
   _____

Score: _____

Name _____ Date _____ Class _____

JOB 76: Using a Corner Pole

OBJECTIVE: After you have completed this job, you will be able to use a corner pole for a particular job using the proper technique.

TEXTBOOK REFERENCE: Study the appropriate sections in Chapter 12, *Laying Brick,* and Chapter 13, *Laying Block,* before you begin this job.

Equipment

To complete this job, you will need the following tools and materials:

- Mason's tools
- Chalk line
- Mortar
- Supply of bricks
- 2 corner poles and braces
- Pair of line blocks
- Mason's line

Recommended Procedure

A corner pole may be used instead of building leads when building a brick or block wall. Corner poles may be purchased or built by masons. A sturdy aluminum corner pole with attached bracing is recommended.

In this job, you will use a corner pole to lay a short (10 bricks long) single-wythe brick wall in running bond to a height of 24″.

1. Collect the tools and materials needed for this job and arrange your work area. Completed ☐
2. Snap a chalk line about 10′ long where the wall is to be located. Completed ☐
3. Set up the corner poles at each end of the chalk line. Plumb and brace them in proper relation to the chalk line. Completed ☐
4. Lay out the first course in dry bond to mark the proper spacing. Completed ☐
5. Attach the line blocks and mason's line to the corner poles at the proper height of the first course. Completed ☐
6. Lay the first course of bricks to the line. Completed ☐
7. Move the line up to the next course height and lay the second course of masonry. Completed ☐
8. Continue the process until the wall is 24″ high. Completed ☐
9. Tool the joints as the mortar hardens sufficiently. Brush the finished wall. Completed ☐
10. Clean up the area and return the tools and materials to their proper places. Completed ☐

Instructor's Initials: _____

Date: _____

Job 76 Review

After completing this job successfully, answer the following questions.

1. What are the advantages of using a corner pole over building leads?

   _____
   _____
   _____
   _____

2. What is a disadvantage of using a corner pole over building leads?

   _____
   _____
   _____

3. What situation is *not* suitable for the use of corner poles?

   _____
   _____
   _____

4. What type of situation is highly suitable for the use of corner poles?

   _____
   _____
   _____

   Score: _____